SpringerBriefs in Applied Sciences and Technology

SpringerBriefs present concise summaries of cutting-edge research and practical applications across a wide spectrum of fields. Featuring compact volumes of 50 to 125 pages, the series covers a range of content from professional to academic.

Typical publications can be:

- A timely report of state-of-the art methods
- An introduction to or a manual for the application of mathematical or computer techniques
- A bridge between new research results, as published in journal articles
- A snapshot of a hot or emerging topic
- An in-depth case study
- A presentation of core concepts that students must understand in order to make independent contributions

SpringerBriefs are characterized by fast, global electronic dissemination, standard publishing contracts, standardized manuscript preparation and formatting guidelines, and expedited production schedules.

On the one hand, **SpringerBriefs in Applied Sciences and Technology** are devoted to the publication of fundamentals and applications within the different classical engineering disciplines as well as in interdisciplinary fields that recently emerged between these areas. On the other hand, as the boundary separating fundamental research and applied technology is more and more dissolving, this series is particularly open to trans-disciplinary topics between fundamental science and engineering.

Indexed by EI-Compendex, SCOPUS and Springerlink.

More information about this series at http://www.springer.com/series/8884

Alexander Zhivov · Rüdiger Lohse

Deep Energy Retrofit—A Guide for Decision Makers

EBC
Energy in Buildings and
Communities Programme

Springer

Alexander Zhivov
US Army Corps of Engineers
Engineer Research and Development
Center, Construction Engineering Research
Laboratory
Champaign, IL, USA

Rüdiger Lohse
Klimaschutz und Energieagentur
Linkenheim-Hochstetten
Baden-Württemberg, Germany

ISSN 2191-530X ISSN 2191-5318 (electronic)
SpringerBriefs in Applied Sciences and Technology
ISBN 978-3-030-66210-3 ISBN 978-3-030-66211-0 (eBook)
https://doi.org/10.1007/978-3-030-66211-0

This Springer imprint is published by the registered company Springer Nature Switzerland AG
The registered company address is: Gewerbestrasse 11, 6330 Cham, Switzerland

EBC is a programme of the International Energy Agency (IEA)

EBC is a programme of the International Energy Agency (IEA)

Preface

The International Energy Agency

The International Energy Agency (IEA) was established in 1974 within the framework of the Organization for Economic Cooperation and Development (OECD) to implement an international energy program. A basic aim of the IEA is to foster international cooperation among the 29 IEA participating countries and to increase energy security through energy research, development, and demonstration in the fields of technologies for energy efficiency and renewable energy sources.

The IEA Energy in Buildings and Communities Programme

The IEA coordinates international energy research and development (R&D) activities through a comprehensive portfolio of Technology Collaboration Programmes. The mission of the IEA Energy in Buildings and Communities (IEA EBC) Programme is to develop and facilitate the integration of technologies and processes for energy efficiency and conservation into healthy, low emission, and sustainable buildings and communities through innovation and research. (Until March 2013, the IEA EBC Programme was known as the IEA Energy in Buildings and Community Systems Programme, ECBCS.)

The R&D strategies of the IEA EBC Programme are derived from research drivers, National Programmes within IEA countries, and the IEA Future Buildings Forum Think Tank Workshops. These R&D strategies aim to exploit technological opportunities to save energy in the buildings sector, and to remove technical obstacles to market penetration of new energy-efficient technologies. The R&D strategies apply to residential, commercial, office buildings, and community systems, and will impact the building industry in five areas of focus for R&D activities:

- Integrated planning and building design.
- Building energy systems.

- Building envelope.
- Community-scale methods.
- Real building energy use.

The Executive Committee

Overall control of the IEA EBC Programme is maintained by an Executive Committee, which not only monitors existing projects, but also identifies new strategic areas in which collaborative efforts may be beneficial. As the program is based on a contract with the IEA, the projects are legally established as annexes to the IEA EBC Implementing Agreement. At the present time, the following projects have been initiated by the IEA EBC Executive Committee, with completed projects identified by (*):

Annex 1: Load Energy Determination of Buildings (*)
Annex 2: Ekistics and Advanced Community Energy Systems (*)
Annex 3: Energy Conservation in Residential Buildings (*)
Annex 4: Glasgow Commercial Building Monitoring (*)
Annex 5: Air Infiltration and Ventilation Centre
Annex 6: Energy Systems and Design of Communities (*)
Annex 7: Local Government Energy Planning (*)
Annex 8: Inhabitants Behaviour with Regard to Ventilation (*)
Annex 9: Minimum Ventilation Rates (*)
Annex 10: Building HVAC System Simulation (*)
Annex 11: Energy Auditing (*)
Annex 12: Windows and Fenestration (*)
Annex 13: Energy Management in Hospitals (*)
Annex 14: Condensation and Energy (*)
Annex 15: Energy Efficiency in Schools (*)
Annex 16: BEMS 1- User Interfaces and System Integration (*)
Annex 17: BEMS 2- Evaluation and Emulation Techniques (*)
Annex 18: Demand Controlled Ventilation Systems (*)
Annex 19: Low Slope Roof Systems (*)
Annex 20: Air Flow Patterns within Buildings (*)
Annex 21: Thermal Modelling (*)
Annex 22: Energy Efficient Communities (*)
Annex 23: Multi Zone Air Flow Modelling (COMIS) (*)
Annex 24: Heat, Air and Moisture Transfer in Envelopes (*)
Annex 25: Real time HVAC Simulation (*)
Annex 26: Energy Efficient Ventilation of Large Enclosures (*)
Annex 27: Evaluation and Demonstration of Domestic Ventilation Systems (*)
Annex 28: Low Energy Cooling Systems (*)
Annex 29: Daylight in Buildings (*)

Annex 30: Bringing Simulation to Application (*)
Annex 31: Energy-Related Environmental Impact of Buildings (*)
Annex 32: Integral Building Envelope Performance Assessment (*)
Annex 33: Advanced Local Energy Planning (*)
Annex 34: Computer-Aided Evaluation of HVAC System Performance (*)
Annex 35: Design of Energy Efficient Hybrid Ventilation (HYBVENT) (*)
Annex 36: Retrofitting of Educational Buildings (*)
Annex 37: Low Exergy Systems for Heating and Cooling of Buildings (LowEx) (*)
Annex 38: Solar Sustainable Housing (*)
Annex 39: High Performance Insulation Systems (*)
Annex 40: Building Commissioning to Improve Energy Performance (*)
Annex 41: Whole Building Heat, Air and Moisture Response (MOIST-ENG) (*)
Annex 42: The Simulation of Building-Integrated Fuel Cell and Other Cogeneration Systems (FC+COGEN-SIM) (*)
Annex 43: Testing and Validation of Building Energy Simulation Tools (*)
Annex 44: Integrating Environmentally Responsive Elements in Buildings (*)
Annex 45: Energy Efficient Electric Lighting for Buildings (*)
Annex 46: Holistic Assessment Tool-kit on Energy Efficient Retrofit Measures for Government Buildings (EnERGo) (*)
Annex 47: Cost-Effective Commissioning for Existing and Low Energy Buildings (*)
Annex 48: Heat Pumping and Reversible Air Conditioning (*)
Annex 49: Low Exergy Systems for High Performance Buildings and Communities (*)
Annex 50: Prefabricated Systems for Low Energy Renovation of Residential Buildings (*)
Annex 51: Energy Efficient Communities (*)
Annex 52: Towards Net Zero Energy Solar Buildings (*)
Annex 53: Total Energy Use in Buildings: Analysis and Evaluation Methods (*)
Annex 54: Integration of Micro-Generation and Related Energy Technologies in Buildings (*)
Annex 55: Reliability of Energy Efficient Building Retrofitting - Probability Assessment of Performance and Cost (RAP-RETRO) (*)
Annex 56: Cost Effective Energy and CO2 Emissions Optimization in Building Renovation
Annex 57: Evaluation of Embodied Energy and CO2 Equivalent Emissions for Building Construction (*)
Annex 58: Reliable Building Energy Performance Characterization Based on Full Scale Dynamic Measurements (*)
Annex 59: High Temperature Cooling and Low Temperature Heating in Buildings (*)
Annex 60: New Generation Computational Tools for Building and Community Energy Systems
Annex 61: Business and Technical Concepts for Deep Energy Retrofit of Public Buildings

Annex 62: Ventilative Cooling
Annex 63: Implementation of Energy Strategies in Communities
Annex 64: LowEx Communities—Optimized Performance of Energy Supply Systems with Exergy Principles
Annex 65: Long-Term Performance of Super-Insulating Materials in Building Components and Systems
Annex 66: Definition and Simulation of Occupant Behavior in Buildings
Annex 67: Energy Flexible Buildings
Annex 68: Indoor Air Quality Design and Control in Low Energy Residential Buildings
Annex 69: Strategy and Practice of Adaptive Thermal Comfort in Low Energy Buildings
Annex 70: Energy Epidemiology: Analysis of Real Building Energy Use at Scale
Annex 71: Building Energy Performance Assessment Based on In-situ Measurements
Annex 72: Assessing Life Cycle Related Environmental Impacts Caused by Buildings
Annex 73: Towards Net Zero Energy Public Communities
Annex 74: Energy Endeavour
Annex 75: Cost-effective Building Renovation at District Level Combining Energy Efficiency and Renewables

Working Group—Energy Efficiency in Educational Buildings (*)
Working Group—Indicators of Energy Efficiency in Cold Climate Buildings (*)
Working Group—Annex 36 Extension: The Energy Concept Adviser (*)
Working Group—HVAC Energy Calculation Methodologies for Non-residential Buildings.

Champaign, USA — Alexander Zhivov
Linkenheim-Hochstetten, Germany — Rüdiger Lohse

Acknowledgements

The "Deep Energy Retrofit—A Guide for Decision Makers" was developed within the IEA EBC Program Annex 61, "Business and technical Concepts for Deep Energy Retrofit of Public Buildings" as the result of a joint effort by researchers and practitioners from Austria, Canada, China, Denmark, Estonia, Germany, Latvia, UK, and the United States. The authors express their appreciation to the many international contributors and organizations whose volunteer efforts contributed to development of a series of Guides developed under this project. Special gratitude to the ASHRAE and its Technical Committees TC 7.6 "Building Energy Performance," TC 5.2 "Duct Design," and a Standing Standard Project Committee 90.1 for providing a platform for public discussion of the project progress and to Chartered Institution of Building Services Engineers (CIBSE) for sharing its technical information and providing valuable input into these documents. The authors would like to personally thank the members of the EBC Program Executive Committee for their directions, guidance, and support to the project. Special appreciation to the following reviewers, who provided their valuable comments and suggested improvements to this Guide: Conny Rolén (the Swedish Research Council Formas), Dr. Seung-eon Lee (Korea Institute of Construction Technology) and Michael Donn (Victoria University of Wellington), Margaret Simmons, (U.S. Army Corps of Engineers), and Kinga Porst Hydras (U.S. General Services Administration). The authors gratefully acknowledge William J. Wolfe, Writer-Editor, ERDC-ITL for his help in coordinating the preparation of this document. The authors would like to acknowledge the financial and other support from the following: the Office of the Assistant Secretary of the US Army for Installations, Energy, and Environment; US Army Corps of Engineers; US Department of Energy (Federal Energy Management Program); The Austrian Ministry for Transport, Innovation and Technology; the EUDP-programme of the Danish Energy Agency and Bundesministerium für Wirtschaft und Energie, Germany.

Contents

Acronyms and Abbreviations

AEE	Austrian Energy and Environment
ANSI	American National Standards Institute
ARRA	American Recovery and Reinvestment Act
ASHRAE	American Society of Heating, Refrigerating, and Air-Conditioning Engineers
ASTM	American Society for Testing and Materials
BEMS	Building Energy Management System
CBD	Central Business District
CERL	Construction Engineering Research Laboratory
CERT	Centre for Economic Reform and Transformation
CFR	Code of the Federal Regulations
CIBSE	The Chartered Institution of Building Services Engineers
COGEN	Cogeneration
COGEN-SIM	The Simulation of Building-Integrated Fuel Cell and Other Cogeneration Systems
COMIS	Conjunction of Multizone Infiltration Specialists
CRD	Capital Requirement Directive
CRR	Capital Requirement Regulation
DER	Deep Energy Retrofit
DEROM	Deep Energy Retrofit Optimization Model
DIN	Deutsches Institut für Normung [the German national standards organization]
DOAS	Dedicated Outdoor Air Supply
DoD	US Department of Defense
DOE	US Department of Energy
EBC	Energy in Buildings and Communities Programme
ECBCS	Energy Conservation in Buildings and Community Systems
ECM	Energy Conservation Measure
EDF	Environmental Defense Fund
EE	Energy Efficiency
EEFIG	Energy Efficiency Financial Institutions Group

EEM	Energy Efficiency Measure
EISA	US Energy Independence and Security Act of 2007
EO	Executive Order
EPC	Energy Performance Contract
ERDC	US Army Engineer Research and Development Center
ESCO	Energy Service Company
ESPC	Energy Savings Performance Contract
EU	European Union
EUDP	Energy Technology Development and Demonstration Program
EUI	Energy Use Intensity
FAR	Federal Acquisition Regulation
FC	Fuel Cell
GC	General Contractor
GESP	Guaranteed Energy Savings Program
GSA	General Services Administration
HR	Heat Recovery
HVAC	Heating, Ventilating, and Air-Conditioning
HYBVENT	Hybrid Ventilation
IAQ	Indoor Air Quality
IEA	International Energy Agency
IPCC	Intergovernmental Panel on Climate Change
IPD	Investment Property Databank
ISBN	International Standard Book Number
ISSN	International Standard Serial Number
IWU	Institut für Wohnung und Umwelt (Institute for Housing and Environment)
JOSRE	Journal of Sustainable Real Estate
KfW	Kreditanstalt für Wiederaufbau (Reconstruction Credit Institute)
LCC	Life Cycle Cost
LCCA	Life-Cycle Cost Analysis
LCP	Least-Cost Planning
LED	Light Emitting Diode
LEED	Leadership in Energy and Environmental Design
M&V	Measurement and Verification
MiFID	Markets in Financial Instruments Directive
NIBS	National Institute of Building Sciences
NPV	Net Present Value
NZE	Net Zero Energy
OBR	On-Bill Repayment (model)
OECD	Organization for Economic Co-operation and Development
OPR	Owner's Project Requirements
PACE	Property Assessed Clean Energy
PDF	Portable Document Format
PPA	Power Purchase Agreement
PV	PhotoVoltaic

QA	Quality Assurance
R&D	Research and Development
RAP-RETRO	Reliability of Energy Efficient Building Retrofitting - Probability Assessment of Performance & Cost
RE	Renewable Energy
RICS	Royal Institution of Chartered Surveyors
RMI	Rocky Mountain Institute
SHGC	Solar Heat Gain Coefficient
SOW	Statement of Work
SRM	Sustainment, Restoration, and Modernization
TC	Technical Committee
UK	United Kingdom
US	United States
USA	United States of America
USACE	US Army Corps of Engineers
USC	US Code
USD	US dollars
ZVEI	Zentralverband Elektrotechnik- und Elektronikindus (German Central Association of Electrical and Electronic Industry)

List of Figures

List of Tables

Chapter 1
Introduction

Many governments worldwide are setting more stringent targets for reduction in energy use in government/public buildings. Buildings constructed more than 10 years ago account for a major share of energy used by the building stock. However, the funding and "know-how" (applied knowledge) available for owner-directed energy retrofit projects have not kept pace with new requirements. With typical retrofit projects, reduction of energy use varies between 10 and 20%, while actual executed renovation projects show that energy use reduction can exceed 50%, and can cost-effectively achieve the Passive House standard or even approach net zero-energy status (EBC Annex 61 2017a, Hermelink and Müller 2010; NBI 2014; RICS 2013; Shonder and Nasseri 2015; Miller and Higgins 2015; Emmerich et al. 2011).

Building energy efficiency (EE) ranks first in approaches with resource efficiency potential with a total resource benefit of approximately $700 billion until 2030. EE is by far the cheapest way to cut CO_2 emissions (McKinsey 2011, IPCC 2007). However, according to an IEA study (IEA 2014a), more than 80% of savings potential in building sector remains untapped. Thus, the share of deployed EE in the building sector is lower than in the Industry, Transport, and Energy generation sectors. Estimates for the deep renovation potentials show: €600-900bn investment potential, €1000-1300bn savings potential, 70% energy-saving potential, and 90% CO_2 reduction potential.

The five key elements of a long-term DER strategy (adapted from IEA 2014b) are as follows:

1. ***Rollout of DER system approaches***. Develop and replicate cost-effective deep energy renovation as part of normal building renovation activity, and make sure that the outcomes of these refurbishments are measured, verified, and evaluated.
2. ***Monitoring and verification***: Set up consistent measurement and verification processes that help to build up a reliable data basis for DER projects. This should be done when evaluated energy performance data adjustments can be initiated,

A. Zhivov and R. Lohse, *Deep Energy Retrofit—A Guide for Decision Makers*,
SpringerBriefs in Applied Sciences and Technology,
https://doi.org/10.1007/978-3-030-66211-0_1

for example, when retro-commissioning before issuing an Energy Performance Certificate every 10 years.

3. ***DER energy performance contracting***. To achieve a life-cycle, cost-neutral approach, both energy and non-energy-related benefits must be quantitatively valued. The private sector invests in modest savings; public policy with its influence and funding can help drive DER.
4. ***Avoid Staging and "Cream-Skimming" in Building Refurbishments***: The overwhelming number of building renovations results in modest energy savings and can be categorized as "shallow refurbishments" (not DER). Shallow refurbishments, especially heating, ventilating, and air-conditioning (HVAC) replacement, largely ignore many missed opportunities if envelope improvements, such as facade upgrade, or roof or window replacement, are not undertaken. "Cream-skimming" the HVAC and other shorter term options will make future investments for remaining items even less appealing since the shortest term investment would have already ready been done. Any kind of building strategy must target such crucial decision points and steer decision-makers toward a "whole-building" approach.
5. ***Provide incentives for DER***: Incentives such as access to public subsidies should be limited to DER projects that target ambitious levels of energy performance that follow approaches of 50% energy use reduction for renovation.
6. **Collect data on DER projects.** To boost DER in buildings, the reliable data must be collected and distributed among the building owners, funding institutes, and energy service companies.

References

A. Hermelink, A. Müller, Economics of deep renovation. European Insulation Manufacturers Association. Project # PDEMDE 101646 (2010)

A. Miller, C. Higgins, Deep energy savings in existing buildings. ASHRAE Transactions, Volume 121, Part 2 (2015)

EBC Annex 61, Deep Energy Retrofit Case Studies. Annex 61 Subtask A report (2017). www.iea-annex61.org

J. Shonder, C. Nasseri, Achieving deeper energy savings in federal energy performance contracts. ASHRAE Transactions, Volume 121, Part 2 (2015)

NBI. 2014. New Buildings Institute. http://buildings.newbuildings.org

RICS, *Sustainable construction: realizing the opportunities for built environment professionals* (RICS Europe, Brussels, 2013)

W. Emmerich, A. Georgescu, M. Ginter, H. Garrecht, J. Huber, O. Hildebrand, A. Koenig, M. Laidig, E. Gruber, R. Jank, *EnSan Projekt Karlsruhe Görderlerstraße Garrecht* (Volkswohnung, Karlsruhe, Germany, 2011)

Chapter 2
Deep Energy Retrofit in Public Buildings—EBC Annex 61 Approach

Research under the IEA EBC Annex 61 has been conducted with a goal of providing a framework, selected tools, and guidelines to significantly reduce energy use (by more than 50%) in government and public buildings constructed before the 1980s with low internal loads (e.g., office buildings, dormitories, barracks, public housing, and educational buildings) undergoing major renovation.

Best practices from Europe (Austria, Denmark, Estonia, Germany, Ireland, Latvia, Montenegro, The Netherlands, United Kingdom) and the United States have been studied and 26 examples of implemented retrofit projects, in which site energy use has been reduced by 50% or better compared to pre-renovation base line, have been documented in the "Deep Energy Retrofit—Case Studies" report (EBC Annex 61 2017a). These case studies were analyzed with respect to energy use before and after renovation, reasons for undertaking the renovation, co-benefits achieved, resulting cost-effectiveness, and the business models followed. Finally, the lessons learned were compiled and compared.

A list of core energy efficiency technologies (Table 2.1) was generated from the results of case studies, from surveys and discussions conducted at the American Society of Heating, Refrigerating and Air-Conditioning Engineers (ASHRAE) Technical Committee (TC) 7.6 "Public Buildings" working group meetings in 2013 and 2014, and from previous experience and research conducted by the EBC Annex 61 team members.

These technologies, when applied together (as a bundle), will reduce the total building site energy use by about 50% (including plug loads). Technical characteristics of these building envelope-related technologies grouped into a "core technologies bundle" have been studied through modeling and life-cycle cost (LCC) analysis for representative national climate conditions and presented in the "Deep Energy Retrofit—A Guide to Achieving Significant Energy Use Reduction with Major Renovation Projects" (EBC Annex 61 2017b) or "DER Technical Guide" for short.

A. Zhivov and R. Lohse, *Deep Energy Retrofit—A Guide for Decision Makers*,
SpringerBriefs in Applied Sciences and Technology,
https://doi.org/10.1007/978-3-030-66211-0_2

Table 2.1 Core technologies bundles for DER

Category	Name
Building envelope	Roof insulation
	Wall insulation
	Slab insulation
	Windows
	Doors
	Thermal bridges remediation
	Airtightness
	Vapor barrier
	Building envelope quality assurance (QA)
Lighting and electrical systems	Lighting design based on LED technologies, daylight and motion controls
HVAC	High-performance motors, fans, furnaces, chillers, boilers, etc.
	Dedicated outdoor air system (DOAS)
	Heat recovery (HR) (dry and wet) with efficiency >70%
	Duct insulation
	Duct airtightness
	Pipe insulation

The Guide provides examples of "best practices" that illustrate optimal methods of applying these technologies in different construction situations. The "Deep Energy Retrofit Business Guide" resulted from the Subtask B to examine facilitation and implementation of DER projects from the business perspective. The evaluation of DER case studies indicates that cost-effectiveness and availability of funding are the most relevant decision-making criteria to initiate a DER concept. The implementation of cost-ineffective DER concepts is one major problem that often leads to shallow refurbishments. Thus, the Business Guide examines strategies to improve the cost-effectiveness of DER projects by reducing the cost of investment and by quantifying energy- and non-energy-related cost savings, i.e., the Multiple Benefits of Energy Efficiency (IEA 2015).

So far the majority of DER projects have only been implemented in a traditionally funded business model, which is limited to the financial, organizational, and technological capacities of public building agencies. The "DER Business Guide" (EBC Annex 61 2017c) provides information on advanced business models that combine financing, implementation, and operation services with performance-related remuneration. The "DER Business Guide" highlights recent efforts to advance the energy performance contracting mechanism and broaden its scope to include ambitious DER projects.

The "DER Business Guide" is illustrated by examples of pilot projects implemented by innovators in Belgium, the United States, Latvia, and Germany.

Seven pilot projects were conducted in the working phase of the EBC Annex 61 with a goal of achieving 50% or more reduction against the energy consumption baseline before the refurbishment:

- Dormitory in Manheim, Germany.
- IWU Office Building in Darmstadt, Germany.
- Almegårds Kaserne Military Barracks in Bornholm, Denmark.
- Presidio Military Barracks in Monterey, California, USA.
- Federal building and courthouse in St. Croix, US Virgin Islands.
- Federal Buildings, Silver Spring, MD, USA.
- Kindergarten, Valga (Estonia).

Some of technical and business concepts developed and described in the EBC Annex 61 Guides and their combinations have been tested and further studied during these pilot projects. Technical concepts implemented in these projects along with the cost information, cost-effectiveness, and lessons learned have been documented in the "DER Pilot Projects" report (EBC Annex 61 2017d).

This document, the "Deep Energy Retrofit—A Guide for Decision Makers," presents the summary of information developed by EBC Annex 61 and designed for building owners, executive decision-makers, energy managers of public organizations, financial institutions, investors, and energy service companies (ESCOs), to help them better understand the opportunities for cost-effective DER.

References

EBC Annex 61, Deep Energy Retrofit Case Studies. Annex 61 Subtask A report. (2017a). www.iea-annex61.org

EBC Annex 61, Deep Energy Retrofit—A Guide to Achieving Significant Energy Use Reduction with Major Renovation Projects. Subtask A report (2017b). www.iea-annex61.org

EBC Annex 61, Deep Energy Retrofit Business Guide. Annex 61 Subtask B report. (2017c). www.iea-annex61.org

EBC Annex 61, Deep Energy Retrofit Pilot Projects. Annex 61 Subtask C report. (2017d). www.iea-annex61.org

Chapter 3
What is Deep Energy Retrofit?

Although the term "Deep Energy Retrofit" is currently widely used, there is no established global definition. Since the energy crisis of the 1970s, energy requirements pertaining to new construction and building renovation worldwide have significantly improved. Since the 1980s, building energy use requirements in the United States have improved by more than 50% (calculated without consideration of plug loads). Furthermore, buildings and building systems degrade over time, with cracks in the building envelope; dirty and leaky ducts; lower efficiencies when HVAC systems are not regularly commissioned; etc. This can reduce their energy performance by at least 10%. It is technically feasible to recoup these inefficiencies and further reduce building energy use by more than 50% by using technologies readily available on the market and by simply adapting current requirements for new buildings to the refurbishment of the existing building stock.

Analysis conducted by the EBC Annex 61 team (EBC Annex 61 2017) shows that a significant number of commercial and public buildings have reduced their energy consumption by more than 50% after renovation, and that some have met the Passive House Institute energy efficiency standard or the net zero-energy state. According to the Global Building Performance Network prognosis (RICS 2013), a DER that follows the most recent and proposed EU guidance can improve the buildings energy performance by at least 80%. Based on these experiences, the IEA EBC Annex 61 team has proposed (Zhivov et al. 2015) the following definition of the Deep Energy Retrofit:

Deep Energy Retrofit (DER) is a major building renovation project in which site energy use intensity (including plug loads) has been reduced by at least 50% from the pre-renovation baseline with a corresponding improvement in indoor environmental quality and comfort.

A. Zhivov and R. Lohse, *Deep Energy Retrofit—A Guide for Decision Makers*,
SpringerBriefs in Applied Sciences and Technology,
https://doi.org/10.1007/978-3-030-66211-0_3

A DER requires a whole-building analysis approach along with an integrative design process. A "whole-building analysis" means that the building is considered as a single, integrated system rather than as a collection of stand-alone systems, such as building envelope, HVAC system, renewable energy system, building operations, etc. The whole-building approach facilitates the identification of synergistic relationships between the component systems. Analyzing systems in isolation does not effectively identify synergies between systems. For example, improving the building envelope, providing solar heat gain control, and improving lighting systems could substantially reduce a building's heating and cooling energy demand. This would, in turn, reduce the required size of duct systems, air-handling units, boilers, and chillers. Likewise, replacing an aging air-handling unit with a smaller, more efficient unit could improve indoor air quality and further reduce energy demand. Such cascading benefits would not be achievable if the building were not analyzed as an integrated whole.

The key to whole-building analysis is the use of an integrated design process. The whole-building analysis differs from a traditional design process in that it brings all relevant disciplines together for an initial charrette-based study of the problem as a whole, based on collaboration and shared information, whereas a more traditional process is based on a linear flow of information passing from one discipline to another.

References

EBC Annex 61, Deep Energy Retrofit—A Guide to Achieving Significant Energy Use Reduction with Major Renovation Projects. Subtask A report (2017). www.iea-annex61.org

RICS, *Sustainable Construction: Realizing the opportunities for built environment professionals* (RICS Europe, Brussels, 2013)

A. Zhivov, R. Lohse, J. Shonder, C. Nasseri, H. Staller, O. Moerck, M. Nokkala, Business and Technical Concepts for Deep Energy Retrofit of Public Buildings. AT-15-009 . ASHRAE Transactions, Volume 121, Part 2 (2015)

Chapter 4
Deep Energy Retrofit Versus Shallow Renovation

Current studies show that the typical approach to refurbishing the building stock is to follow a "shallow renovation" track that focuses on single measures and partial refurbishments, primarily on lighting retrofits, HVAC replacement, and retro-commissioning; and on other ECMs that provide low risk and short payback periods. Such projects rarely include such measures as facade and roof insulation, replacement of windows, remediation of thermal bridges, or significant improvements in building airtightness. In countries with stringent source energy targets for refurbishment projects, building owners tend to choose renewable energy and heating supply solutions over measures that improve energy efficiency, increase insulation, etc.

From the perspective of a public building owner, shallow refurbishments, especially HVAC replacements, offer a large risk for a "missed opportunity" if envelope improvements such as facade upgrade, or roof or window replacement are not undertaken. A combined approach would have allowed a downsizing of the HVAC system due to lower heating and cooling demands, and elimination of perimeter zone conditioning, and would likely have provided improved comfort. More importantly, the findings from the EBC Annex 61 pilot case studies show that a combined bundle of HVAC, thermal envelope, and renewable power and heat supply with individual short- and long-term payback periods are likely to be cost-effective. For the decision-making on the building level, it is necessary to identify cost-effective pathways for DER instead of considering minimum requirements in "shallow refurbishment" approaches.

This is also true on the macroeconomic level. Dynamic simulations (Bettgenhäuser et al. 2014; Nock and Wheelock 2010) have clearly shown that "deep renovation at reasonable speed" is a more promising strategy to reach long-term (2050) climate targets than "shallow renovation at high speed." "Shallow renovation" with very high shares of renewable energy undertaken to achieve source energy targets appears to be 3.5% more expensive.

Currently, many European national implementation strategies recommend that building refurbishment be conducted using a step-by-step or phased approach. While

A. Zhivov and R. Lohse, *Deep Energy Retrofit—A Guide for Decision Makers*, SpringerBriefs in Applied Sciences and Technology,
https://doi.org/10.1007/978-3-030-66211-0_4

limited scope projects with replacement of a single piece of equipment and projects addressing separable sections of the building at different times of the building life are justifiable, implementing different measures using a step-by-step approach will result in increased total investment costs; more complex project planning, contracting, and design; and in higher operation costs due to suboptimal selection of technologies, their characteristics, and sizes for each step. The EBC Annex 61 has established bundles of core energy efficiency technologies and their characteristics (see Section 2), which, when implemented together, result in cost-effective solutions for DER. Sections 7, 8, and 9 describe strategies on how to make DER project cost-effective as well as business models and project financing strategies that provide practical information for building owners with limited funding available for building refurbishment.

References

Bettgenhäuser Kjell, Rolf de Vos, Jan Grözinger, Thomas Boermans, Deep renovation of buildings. An effective way to decrease Europe's energy import dependency. Ecofys project # BUIDE14901. Prepared for Eurima (2014)

Nock, Levin, Clint Wheelock, Energy Efficiency Retrofits for Commercial and Public Buildings. Energy Savings Potential, Retrofit Business Cases, Financing Structures, Policy and Regulatory Factors Demand Drivers by Segment, and Market Forecasts. Pike Research (2010)

Chapter 5
Major Renovation and Deep Energy Retrofit

Typical energy efficiency improvement projects are planned as follows:

- A part of major building renovation.[1]
- A part of minor building renovation.
- Utilities modernization projects.
- Mechanical and electrical equipment/systems replacement.
- System retro-commissioning.

The need to reduce energy consumption is one of the many reasons buildings undergo major renovations. Some of the most common reasons to renovate buildings are to

- Extend the building's useful life with an overhaul of its structure, internal partitions, and systems.
- Repurpose the building (e.g., renovation of old warehouses into apartments).
- Bring the building to new or updated codes such as fire protection.
- Remediate environmental problems (mold and mildew); improve the visual appearance, thermal comfort, or indoor air quality.
- Add to the value with improvements to increase investment (increasing useful space and/or space attractiveness/quality) resulting in a higher sale or lease price.

Timing a DER to coincide with a major renovation is best since during the renovation, the building is typically evacuated and gutted; scaffolding is installed; single-pane and damaged windows are scheduled for replacement; building envelope insulation is replaced and/or upgraded; and most of mechanical, electrical lighting, and

[1] The U.S. Department of Energy (DOE) (DOE 2010) and Europe's Energy Performance of Buildings Directive (EU 2010) define a major building renovation as any renovation where the cost exceeds 25% of the replacement value of the building. EPBD also defines building renovation as a major renovation if more than 25% of the surface of building envelope undergoes renovation. US DOD policy (2013) defines a major renovation project as one in which renovation costs exceed 50% of estimated replacement costs (ERC).

A. Zhivov and R. Lohse, *Deep Energy Retrofit—A Guide for Decision Makers*, SpringerBriefs in Applied Sciences and Technology,
https://doi.org/10.1007/978-3-030-66211-0_5

energy conversion systems (e.g., boiler and chillers) along with connecting ducts, pipes, and wires are replaced. A significant sum of money covering the cost of energy-related scope of the renovation designed to meet minimum energy code (a significant part of the DER) is already budgeted in a major renovation.

References

DOE, Energy efficiency and sustainable design standards for new Federal buildings (2010). www.govpulse.us/entries/2010/05/28/2010-12677/energy-efficiency-and-sustainable-design-standards-for-new-Federal-buildings#idp22682672.Washington, DC: DOE

Energy Performance of buildings Directive 2010/31/EU. http://eur-lex.europa.eu/LexUriServ/LexUriServ.do?uri=OJ:L:2010:153:0013:0035:EN:PDF

Chapter 6
Product Delivery Quality Assurance Process

To increase a building's value and improve its indoor climate and thermal comfort, DER must adopt a QA process that, in addition to conventional understanding of QA, includes

- Formulation of detailed technical specification, e.g., statement of work (SOW) or Owner's Project Requirements (OPR), against which tenders (i.e., bids) will be made, and verification that potential contractors understand these specifications.
- Specification in the SOW/OPR of areas of major concern to be addressed and checked during the bid selection, design, construction, commissioning, and post-occupancy phases.
- Clear delineation of the responsibilities and qualifications of stakeholders in this process.

A DER building project must be properly implemented through all phases to accomplish the goals and achieve the owner's performance targets. This requires a project-specific QA process. A properly implemented DER will increase a building's value, improve its indoor climate and thermal comfort, and meet owner's energy and sustainability goals. DER is best accomplished by adopting a project-specific QA process. The process as described in this document supplements those procedures addressed in current standards and guides that specifically address DER, sustainability, and energy conservation in buildings (NIBS 2012; ASTM 2015; ASHRAE Guideline 0-2013). The QA process for a specific project must be developed to suit the needs and goals of that specific project.

This process is applicable to DER projects using any procurement method, including Design-Bid-Build and Design/Build approaches. For more details, see Appendix B and EBC Annex 61 (2017d).

A. Zhivov and R. Lohse, *Deep Energy Retrofit—A Guide for Decision Makers*, SpringerBriefs in Applied Sciences and Technology,
https://doi.org/10.1007/978-3-030-66211-0_6

References

ASHRAE, The Commissioning Process. Guideline 0-2013. Atlanta, GA: ASHRAE (2013)

ASTM, American Society for Testing and Materials (ASTM). Standard Practice for Building Enclosure Commissioning. E2813-12. ASTM International (2015)

EBC Annex 61, Deep Energy Retrofit—A Guide to Achieving Significant Energy Use Reduction with Major Renovation Projects. Subtask A report (2017). www.iea-annex61.org

NIBS, National Institute of Building Sciences. Building Enclosure Commissioning Process BECx. Guideline 3-2012. Washington, DC: NIBS (2012)

Chapter 7
How to Make DER Cost-Effective?

7.1 Introduction

The scope of a DER project and its attractiveness to investors depends on the project's cost-effectiveness. The standard method to analyze a project's cost-effectiveness is by performing a life-cycle cost analysis (LCCA), which accounts for present and future costs of the project.

An important consideration in LCCA is the selection of the base case scenario, against which cost-effectiveness of DER will be evaluated, not to be confused with the baseline, which is used for benchmarking energy use in the building prior to renovation. Most of the major renovation projects include a scope of work that can be non-energy related and the one that is energy related.

Life-cycle costs typically include the following two categories: investment costs and operational costs. Investment-related costs include costs related to planning, design, purchase, and construction as well capital replacement costs, which are usually incurred when replacing major systems or components. An LCCA also typically includes energy use cost and the cost of operation and maintenance.

Most of major renovation projects include a scope of work, which can be either non-energy related or energy related. A non-energy-related scope of work may include such elements as different construction jobs related to changing floor layouts (e.g., moving/removing internal partitions), adding bathrooms, removing asbestos, adding sprinkler system, etc.

An energy-related scope of work of a major renovation project typically includes replacement of existing mechanical, lighting, and electrical systems, replacement of some or all windows, replacement of existing ductwork and plumbing systems, etc. A major renovation with the energy-related scope of work undertaken to meet current minimum standard requirements will be considered to be a base case for the LCCA.

A. Zhivov and R. Lohse, *Deep Energy Retrofit—A Guide for Decision Makers*, SpringerBriefs in Applied Sciences and Technology,
https://doi.org/10.1007/978-3-030-66211-0_7

The main drivers for improving cost-effectiveness of a DER projects include optimization of investment costs and accounting for all operational costs savings and for additional non-energy-related life-cycle cost benefits.

7.2 Investment Cost Reduction

Energy-related investment costs will usually be higher in a DER compared to the base case. Some energy-related improvements included in a DER, e.g., building envelope insulation and mitigation of thermal bridges, installation of high-performance windows, and airtightening the building envelope, are expensive and are rarely included in the scope of major renovation. However, reduction of heating, cooling, or humidity loads resulting from implementation of these DER measures will result in the need for a smaller and sometimes simpler HVAC system, which will, in turn, reduce both initial investment and capital replacement costs related to these systems. In addition, the energy-related improvements should be assessed using a least-cost-pathway approach.

Timing a DER to coincide with a major renovation will improve the cost-effectiveness by reducing the incremental cost to achieve the DER since the building is typically evacuated and gutted; scaffolding is installed; single-pane and damaged windows are scheduled for replacement; building envelope insulation is replaced and/or upgraded; and most of mechanical, electrical, lighting, and energy conversion systems (e.g., boiler and chillers) along with connecting ducts, pipes, and wires will be replaced. A significant sum of money covering the cost of energy-related scope of the renovation designed to meet minimum energy code is already budgeted in a typical non-energy renovation.

7.3 Planning Cost Reduction

DER planning costs may account for 15–30% of the overall first costs. The "DER Technical Guide" provides valuable information on limited bundle of technologies that allow for 50% or better energy use reduction. Technical characteristics of these technologies (e.g., building envelope insulation values, characteristics of window, level of air tightness, etc.) listed in the DER Technical Guide can be used as a starting point for further project optimization and potentially for reduction in planning costs.

7.4 Quality Assurance

A DER project must be properly implemented through all phases to accomplish the goals and required performance levels of the owner. Like all construction projects, there are many steps, decisions, and operations that require an orderly application, and a subsequent QA process to avoid significant cost increase in the construction and the operation phases. A properly implemented DER will increase a building's value, improve its indoor climate and thermal comfort, and meet the owner's energy and sustainability goals. When established and well understood, the QA process requires minimum or no additional cost.

7.5 Selection and Optimization of the DER Scenario

The energy-related scope of work and specific characteristics of technologies to be used can be selected using energy modeling. The scenarios to be considered may include: DER (50% of energy use reduction compared to the baseline), renovation to the new building standard, and a "dream scenario," which could be Passive House, near-zero energy (NZE), etc. For each scenario, investment costs and operating cost savings are estimated and then compared to the base case scenario. From the cost–benefit analysis provided for each scenario, the decision-maker then selects the scenario to be further fine-tuned using the least-cost-planning approach (see DER Business Guide). A review of completed refurbishment projects shows that the application of least-cost planning (LCP) can improve the cost-effectiveness of DER projects by 5 to 28% (Reinhardt Jank et al. 2017).

7.6 Operating Cost Reduction

Compared to the base case, DER may or may not result in the following operating cost savings:

- Energy use and cost reduction due to improved efficiency of the building and its systems.
- Energy cost reduction due to shifting energy peaks, switching to different fuels (e.g., using cogeneration or tri-generation), or replacing fossil-fuel-based thermal or electrical systems to systems using renewable energy sources.
- Maintenance cost reduction with replacement of worn equipment at the end of its life cycle.
- Maintenance cost reduction due to downsizing of mechanical systems with reduced heating and cooling loads.
- Operation cost reduction using advanced building automation systems.

In some scenarios, energy use may increase compared to the base case due to new indoor air quality or thermal comfort requirements. For example, adding cooling or humidity control requirement for the building undergoing renovation will result in additional energy use for cooling systems. Maintenance costs of some replacement systems may increase due to the complexity of their controls system, but they may also be offset by reduced energy use resulting from more efficient operation of the HVAC system.

7.7 Other Bankable Cost Benefits

Early studies and pilot projects executed around the world by frontrunners indicate that, in addition to traditional areas of operating cost reduction listed above, there are other bankable cost reduction and income-generating opportunities related to DER (described in more detail in Appendix C) that shall be considered in LCCA:

- Improved building durability due to better temperature and humidity control (reduced annual maintenance and repair cost for building envelope and mold mitigation).
- Grants, rebates, and other financial subsidies for energy-efficient and sustainable design (one-time payment to reduce first investment).
- Reduced costs and time associated with accommodating a "churn" of employees in flexible and sustainable work spaces (single or multiple time cost reduction).
- Increased usable space due to downsized and consolidated mechanical equipment (reduced annual maintenance and repair costs and additional income-generating cash flows).
- Increased usable space due to improved thermal comfort in areas close to external walls (additional income-generating cash flows).
- Increased usable space due to thermal insulation and ventilation of the attic space (additional income-generating cash flows).
- Reduced short-term absenteeism due to improved indoor air quality and comfort (additional staff productivity income-generating cash flows).
- Improved workers' productivity due to improved indoor air quality and comfort (additional staff productivity income-generating cash flows).
- Recruiting and retention cost savings through employee satisfaction (additional cost reduction that can be spread over time).
- Additional revenues from the enhanced demand for deep retrofit properties from potential tenants (additional income-generating rental rate cash flow).
- Reduced insurance premiums resulting from building components' replacement and improved protection against losses (additional annual cost reduction).

The direct and indirect cost savings beyond energy cost savings due to DER estimated based on industry reports and studies summarized by the Rocky Mountains Institute (RMI 2015) are presented in Table 7.1.

Analysis conducted under the EBC Annex 61 showed the following life-cycle cost savings:

- Replacement of equipment that is at the end of its life cycle and requires significant maintenance and replacement costs, which can contribute to another 20–30%.

Table 7.1 Direct and indirect cost savings beyond energy cost savings due to DER

Maintenance costs (Fowler et al. 2008; Leonardo Academy, 2008, Aberdeen Group (2010)	↓ **9.0–14%**
Occupational satisfaction GSA (2011)	↑ **27–76%**
Rental premium Eicholtz, Kok & Quigley (2010), Wiley et al. (2011), Fuerst & McAlister (2011) Eicholtz, Kok et al. (2011), Kok et al. (2011), Newel, Kok et al. (2011), Miller, Kok et al. (2011), Pogue et al. (2011), McGraw Hill/Siemens (2012)	↑ **2.1–17%**
Occupancy premium Wiley et al. (2011), Pogue et al. (2011), McGraw Hill/Siemens (2012)	↑ **3.14–18%**
Property sale price premium Eicholtz, Kok & Quigley (2010), Fuerst & McAlister (2011), Eicholtz, Kok et al. (2011), Newel, Kok et al. (2011)	↑ **11.1–26%**
Employee productivity Lawrence Berkeley National Laboratory	↑ **1.0–10%**
Reduced employee sick days Miller, Poque, Gough & Davis (2009), Cushman, Wakefield et al. (2009), Dunckley (2007), City of Seattle (2005), Romm & Browning (1995)	↑ **0–40%**

- Reduced investment costs by sizing all equipment and the execution of DER project in one phase rather than in several consecutive steps can contribute additional 5–10%.
- Improved thermal comfort and indoor air quality (IAQ) resulting in higher staff productivity and reduced absenteeism results in savings comparable to 100–200% of the energy cost savings.
- Increase in usable floor space (close to insulated external walls and advanced windows, reduced leakage through the building envelope) by ~ 10% produces a value comparable to additional 20–50% from energy cost savings.
- The combination of DER with installation of renewable energy technologies eligible for subsidies or rebates can improve overall project cost-effectiveness by adding another 30–50% to energy cost savings.

7.8 Cost-Effectiveness of DER

Based on the above discussion, the cost-effectiveness of a DER project can be evaluated by conducting an LCCA using incremental investment cost increase (ΔC) required to achieve a DER compared to the base case scenario (Fig. 7.1).

In the LCCA, the DER project total cost of the owning, operating, maintaining, and, sometimes, eventually disposing of the building or its systems over a given study period, are compared to the Base Case with all costs adjusted (discounted) to reflect the time value of money.

The study period for an LCCA, which is the time over which the costs and benefits are related to a capital investment decision of interest to the decision-maker, is determined by the investor's time horizon. The study period begins with the base date and includes both the planning and construction period and the relevant service period, which begins with the service date and extends to the end of the study period. All operation-related costs are assumed to be incurred during the service period. In European Union countries, the study period may be relatively short (as long as 10–15 years), while in the United States, Federal Energy Management Program rules in 10 CFR 436 allow the study period to be as long as 25 years.

For a DER to be cost-effective, this delta investment cost increase shall be smaller than the net present value (NPV) of operating and maintenance costs saving combined with the NPV of replacement cost savings and the NPV of other bankable cost reduction and income-generating opportunities related to DER. Since most of the

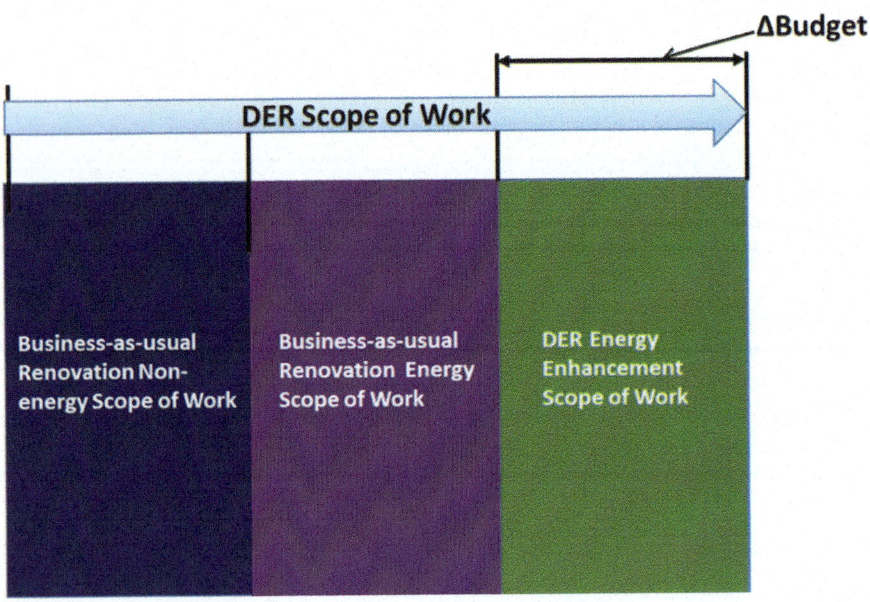

Fig. 7.1 Scope of work of DER project

parameters required for the LCCA differ not only by the individual country, but also within the country (first costs and labor rates, energy rates, life of the project, inflation and discount rates, etc.), the following methodology has been proposed (Zhivov et al. 2015) to evaluate the effectiveness of an LCCA of an integrated energy technology bundle to be used for a DER:

- Step 1. Calculate annual operational costs and income-generating cash flows per DER scenario.
- Step 2. Calculate annual operational costs per base case scenario.
- Step 3. Subtract costs calculated in Step 1 from those calculated in Step 2 and calculate NPV of cost savings over the project life.
- Step 4. The NPV of operational savings and income-generating opportunities can be used to estimate the extent of the budget increase compared to the base case that can be used for energy enhancements with DER compared to building renovation based on minimum energy requirements.

References

Aberdeen Group, Facilities Management: Strategies to Reduce Operating Costs (2010)

P. Eichholtz, K. Nils, J. Quigley, Doing Well by Doing Good? Green Office Buildings. Am. Eco. Rev. **100**, 2492–2509 (December 2010). http://www.aeaweb.org/articles.php?doi=10.1257/aer.100.5.2492

Fowler Kim M, Emily M. Rauch, Assessing Green Building Performance, A Post-Occupancy Evaluation of Twelve GSA Buildings, Pacific Northwest National Laboratory, (July 2008)

GSA, Green Building Performance. A post occupancy evaluation of 22 GSA Buildings. White paper (August 2011)

R. Jank, M. Korb, R. Lohse, *Deep Energy Retrofit Optimization Model (DEROM) Handbook (in German)* (KEA, Karlsruhe, Germany, 2017).

Leonardo Academy, The Economics of LEED for Existing Buildings. Leonardo Academy Inc. (2008, April 21)

A. Zhivov, R. Lohse, J. Shonder, C. Nasseri, H. Staller, O. Moerck, M. Nokkala, Business and technical concepts for deep energy retrofit of public buildings. AT-15–009. ASHRAE Trans. **121**, Part 2 (2015)

Chapter 8
Business Models for DER

This section of the DER Technical Guide describes four business models that are available for energy efficiency retrofit projects to public and commercial sectors. For each business model, a product's value proposition, infrastructure, customers, and finances are described.

8.1 Owner-Directed Model

In this business model, the building owner takes responsibility for the project design, management, and financing of an energy efficiency retrofit. The owner also takes full responsibility (and assumes full liability) for the quality of the project and the economic returns on their investments. The project can be executed using "bid-build" or "design-bid-build" models. The building owner controls contracting, retrofit component selection (and hence the retrofit project price), project management of the work, and is fully liable for the retrofit's subsequent economic performance (i.e., volume of energy required to deliver post-retrofit living conditions) as the financing has recourse only to the owner (possibly secured), but not directly to the retrofit components or to its overall energy performance. By assuming all the components of the retrofit's risk, the building owner is well placed to benefit from any economic outperformance (i.e., when energy prices go up faster than planned) and clearly can benefit directly from a higher grade Energy Performance Certificate and improved acoustics and livability.

For public buildings, this implementation method is limited by the financial, organizational, and technological capacities of the public agencies (Sweatman and Managan 2010).

For example, in the United States, government buildings benefited greatly from appropriations delivered under the American Recovery and Reinvestment Act of 2009 (ARRA). ARRA awarded $5.5 billion to the General Services Administration

A. Zhivov and R. Lohse, *Deep Energy Retrofit—A Guide for Decision Makers*,
SpringerBriefs in Applied Sciences and Technology,
https://doi.org/10.1007/978-3-030-66211-0_8

(GSA) and $7.4 billion to the US Department of Defense (DoD) for the construction and renovation of buildings for energy efficiency improvements and other modernization efforts. ARRA resulted in many impactful projects, including the GSA's net zero-energy retrofit of the Wayne N. Aspinall Federal Building and US Courthouse. However, appropriated funding for building modernization and energy efficiency projects has been less prevalent for US government buildings since ARRA.

8.2 Fixed Repayment

Fixed repayment (primarily used by commercial building owners) is the model in which the upfront capital cost of an energy efficiency retrofit is organized, subsidized, and at times fully provided by either utility or government. An example in the United States is a fixed repayment through a Property Assessed Clean Energy (PACE) program financing mechanism established by a city, county, or Port Authority in the United States. These investments are repaid through monthly, fixed, non-performance-related surcharges.

The "Utility Fixed Repayment" version of this model requires a supportive policy framework to function and the types of legislative changes that regulators have may include (Sweatman and Managan 2010) the following:

- Requirement that electric and gas utilities improve the energy efficiency of their customers by a certain amount each year.
- White certificate programs.
- Decoupling utility profits from the quantity of electricity sold and requirements that utilities invest first in the lowest cost sources of energy.

The "Utility Fixed Repayment" model has several immediate advantages over the owner-directed model:

1. Utility cost of finance, access to funds and available leverage should be considerably better than that achieved by owners under owner-directed model.
2. Friction costs are reduced from the economies of scale created by a utility executing many hundreds or thousands of its individual client retrofits.
3. Customer "ease of execution" is enhanced as execution is streamlined and there is less work for the building's owner than in owner-directed model.
4. Government can use its relationship with the utility sector to align interests and push national energy efficiency targets down to the corporate level through the imposition of standards and markets-based programs like CERT in the UK or the white certificate scheme in Italy.

There are, of course, pros and cons of using energy companies as the main channel for the achievement of government energy efficiency targets. As aggregators, they are the natural partners: energy is their business; they have lots of customers, with access to the energy data required to profile them; and utilities are structured to make major, long-term structural investments in electricity or gas markets. However, without fully decoupling energy suppliers' profitability from the gross amount of energy sold and moving to a smart-grid world where "quality aspects" might dominate, it is hard to see to what extent and for how long energy efficiency can be their top priority.

On-bill repayment (OBR) model is offered by Environmental Defense Fund (EDF) in several states in the United States. It can work for single-family, multi-family, and commercial buildings. It can also work for both tenant-occupied and owner-occupied properties. OBR can accommodate a variety of energy-saving opportunities including equipment purchases, equipment leases, Energy Service Agreements, and Power Purchase Agreements. While on-bill financing refers to programs that use ratepayer, utility shareholder, or public funds, on-bill repayment programs leverage private, third-party capital for financing. Banks, credit unions, or financial institutions provide the loan capital, and loan payments are displayed on utility bills. This approach allows third-party institutions to take care of administrative functions, while utilities need only to process payments. On-bill repayment obligations can use several different financing vehicles, including loans, leases, and power purchase agreements (or PPAs, which serve as agreements to buy and sell energy savings over time).

Property Assessed Clean Energy (PACE) is a modification of fixed repayment model financing mechanism that enables low-cost, long-term funding for energy efficiency, renewable energy, and water conservation projects. PACE financing is repaid as an assessment on the property's regular tax bill, and is processed the same way as other local public benefit assessments (e.g., sidewalks, sewers) have been for decades. Depending on local legislation, PACE can be used for commercial, nonprofit, and residential properties. PACE can cover 100% of a project's hard and soft costs with financing terms up to 20 years. It can be combined with utility, local, and Federal incentive programs. Energy projects are permanently affixed to a property's tax bill, stays with the building on sale, and is easy to share with tenants. PACE is a popular model in the United States, but has not been widely implemented in the EU.

The PACE financing model (Fig. 8.1) works as follows:

1. City, county, or Port Authority creates financing district.
2. Property owner voluntarily applies for financing (which is typically combined with utility or other incentive programs).
3. Proceeds from financing are provided to property owner to pay for project.
4. Property owner installs projects and repays the loan through property tax bills (up to 20 years).

8.3 Energy Performance Contracting Model

Energy-saving performance contracting (ESPC) is the model in which an energy efficiency retrofit provider designs a retrofit, finances it, and is repaid only through the energy savings, therefore assuming the responsibility for the economic success and quality of the retrofit. Recently, in Europe and the United States, advanced ESPC

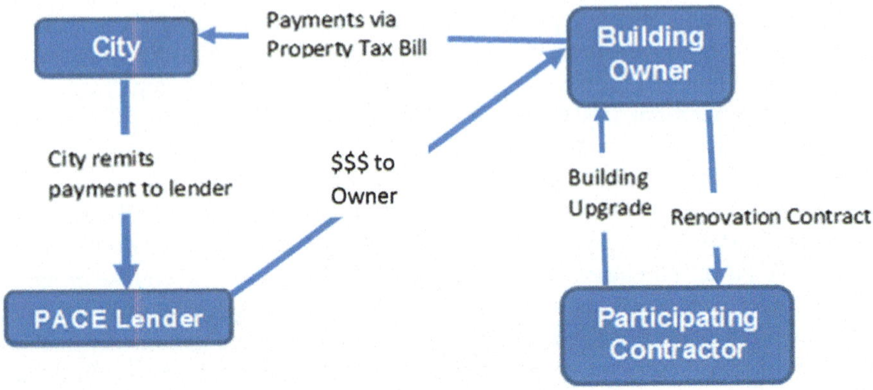

Fig. 8.1 PACE financing (www.pacenation.us/commercial-pace)

schemes are in use that also account for non-energetic life-cycle costs. The "pay-as-you-save" remuneration model of the ESPC allows public entities to increase their funding sources without additional appropriations from fiscal budget legislation. Energy performance contracting is typically delivered in the form of ESPCs. In the United States, ESPCs and Guaranteed Energy Savings Programs (GESPs) [NEA] allow Federal agencies to deliver energy savings without special appropriations from Congress. Project costs are financed by a third party and are paid back over time based on the expected energy savings of the project. The structure of ESPCs is explored further in the Introduction and in Fig. 8.4. In the United States, performance contracts have been essential to enabling agencies to achieve the aggressive energy efficiency targets laid out in Executive Order (EO) 13693 and in the Energy Independence and Security Act (EISA 2007); they will be essential to continued progress as buildings move toward greater levels of efficiency and net zero energy. In the EU, the European Building Performance Directive (EBPD 2016) recommends the use of ESPCs to increase the refurbishment rate. In many of the EU states, however, ESPCs are considered to be third-party financed debt that requires approval from public fiscal controlling institutions. In five EU countries, the ESPC market framework is considered mature and well established (ESCO 2015 Report, Bartoldi et al.).

The energy performance model has most of its application and success in large retrofit projects. In the United States, in 2008, ESCOs have received 88% of their revenue from government buildings and public housing projects, 6% from utility programs, and only 7% from private sector commercial and industrial projects. In EU countries with a mature ESPC market, the public sector is less dominant (50–60% of ESCO revenues). The growth of the energy performance model into smaller segments of the market has been limited because estimating precise energy savings and measuring them in real time to generate bills that are "guaranteed" to save money carries many transaction costs that often cannot be justified for small projects (Sweatman and Managan. 2010). In Germany, specific small- and medium-sized ESCO (SMESCO) structures have been developed to provide ESPC services to medium-sized and small enterprises and buildings (Contracting4KMU, Lohse et al. 2017).

8.4 Blended Funding (Public and Private Combined Funding)

This is a relatively new model in which applying appropriated funding to ESPC projects as a one-time payment (attributed to a cost avoidance) can improve the economics by reducing the total cost to be financed (Junglas et al. 2017). This model allows the project to include longer payback measures, increasing the amount of energy savings and infrastructure renewal that an ESPC would not be able to achieve without this one-time payment.

In the United States for some government agencies like the DoD, this appropriated funding must be designated solely for energy-related projects before being used as supplementary ESPC funding. There is a long history of agencies using appropriated funds, including energy-designated DoD sustainment, restoration, and modernization (SRM) funds, as one-time payments in ESPC projects. There is often a strong argument for applying funds designated for non-energy projects as a one-time payment for an ESPC project to drive greater value, but the legal limitations of combined funding models must be considered.

To maximize the value of DERs, agencies need to both understand the opportunity of pursuing a DER with combined funding sources, and be prepared to act when the timing is right. Developing an energy master plan developed by an unbiased third party is the key first step to understanding the opportunities that a site may offer, and that can inform the need for appropriated funding and potential ESPC projects over time. This energy master plan should be closely coordinated with an energy capital investment plan so that an agency can be prepared to execute and fund energy-related projects appropriately as funding becomes available. Additionally, the energy master plan should remain flexible to pursue combined funding projects as energy-related funds become available.

Precedent for combined funding. There is currently a precedent for combining ESPCs with appropriated funding in situations where that funding has been specifically designated for "related" projects, where the appropriated funds are intended for energy-related projects. The US Department of Energy (DOE) Federal Energy Management Program (FEMP) Guidelines Regarding One-Time Savings Payments and One-Time Savings in ESPCs or cost avoidance in ESPC. These guidelines explain how appropriated funds can be applied to an ESPC. The guidelines apply to projects that are solicited and awarded as an ESPC. The law, 42 USC 8287, has a provision that allows some appropriated funding to be applied to an ESPC. This enabling legislation provides that ESPCs are for the purpose of "achieving energy savings and benefits ancillary to that purpose."

It is imperative that the appropriated funds that are going to be applied to an ESPC are directly related to the energy measures being executed by the ESCO. For example, if an agency had funding available that was intended to replace existing single-pane windows with slightly more efficient double-pane windows, an ESCO, as part of an upcoming ESPC, could finance the incremental cost of more advanced triple-pane windows that will further reduce building loads. The appropriated funding for the original window replacement could be applied to the ESPC as a one-time payment, which would drive greater value from the window replacement through added energy savings and overall project cost-effectiveness. If this project is timed with the trigger of central HVAC system replacement, the reduced heating and cooling loads from the triple-pane windows could allow a less expensive, lower capacity HVAC system replacement. These synergistic approaches are what enable 50% savings achieved in deep retrofits (Zhivov et al. 2015, Shonder and Nasseri 2015).

*Timing is key***.** The alignment of the work being performed by the ESCO with the arrival of appropriated funding that could be applied to the ESPC is critical when evaluating the applicability of those funds to the ESPC. One key challenge faced by the US Army is that an installation is not always certain of which appropriations will be approved until Congress takes action to approve budgets, which can take place 3 to 6 months into a fiscal year. If an ESCO performing work at a given installation is made aware of the energy-related items included in the budget, the ESCO could more deliberately evaluate additional ECMs that could be implemented if the budget is approved and the installation gets the funding. However, there is added development risk for the ESCO and schedule risk for both parties if the ESPC needs to move forward before the funding is received. If the funding does not come through, the applicable ECMs would need to be removed from the project if they could not be paid for as part of the stand-alone ESPC.

Challenges in combining energy and non-energy projects. While a combined funding approach can deliver deeper savings on limited budgets, there are several barriers that prevent broad implementation of this model for the US Federal Government agencies. These limitations do not apply to other cases including state and city government projects. In Federal contracts, ESPCs can only be paid from the savings that are generated from work that is executed as part of the ESPC. When an installation receives appropriated funding for an SRM project, then that project is supposed to be solicited based on the rules in the Federal Acquisition Regulation (FAR). This process can, but does not currently, consider the potential to combine an ESPC effort with the SRM "funding" that could be used for "related" (energy-related) projects. If there is no relationship between the ESPC projects and the "funded" project, the FAR would prevail and the non-energy-related scope would need to be solicited separately from the ESPC efforts. In the combined funding Model #1 illustrated by Fig. 8.2, the general contractor (GC) constructs the entire project, but energy-related portion is implemented under a subcontract with ESCO. GC has two managers (the government customer and the ESCO), but the government customer is ultimately in charge of entire project.

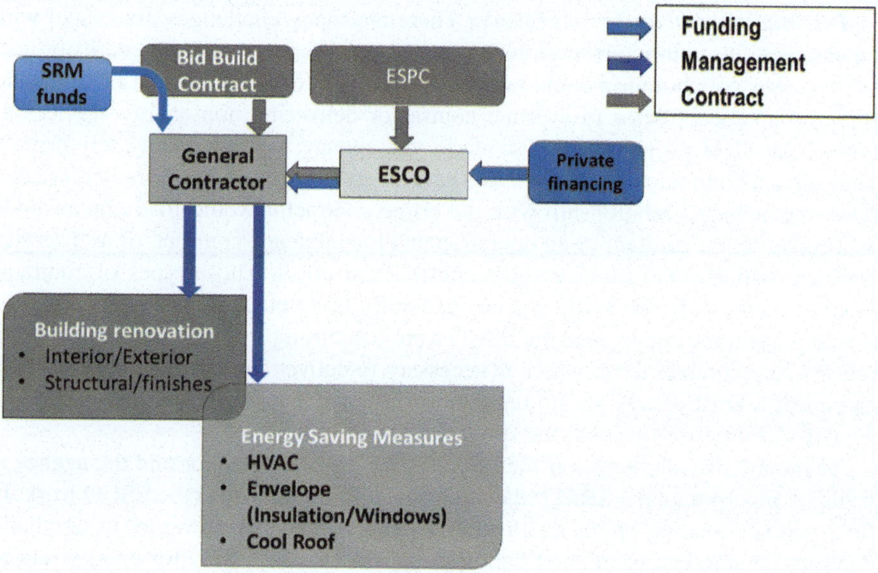

Fig. 8.2 Schematic of the combined funding Model #1

Soliciting non-energy-related scope separately from the ESPC efforts would significantly complicate the project's efforts. From a logistical standpoint, having two or more contractors onsite implementing closely intertwined scopes adds significant complexity to project implementation. Client teams would need to coordinate two contractors with different contracts, schedules, sub-contractors, and scopes to work together in the same space, at the same time, without adversely impacting the project as a whole.

The major legal limitation is not necessarily identifying what scope can be performed by the ESCO under an ESPC. The legal limitation relates to whether or not an agency can advertise a "funded" project as an ESPC, since by law, ESPCs are third-party-financed arrangements. Generally, an ESCO may only perform energy- or water-related conservation measures and related ancillary construction (such as concrete pads under and enclosures around equipment) and operations and maintenance work. If a "funded project" is solicited to an ESCO group, it is likely that the contract community that normally bids those types of projects would protest that the work is not ESPC work. However, current rules would allow an ESCO that is performing related work to use funding as a one-time payment (for agency cost avoidance) if the funding becomes available to use during the right stage of ESPC development. However, the challenge of timing remains significant. Early communication and awareness at an agency or installation regarding projects that could build on each other to achieve savings is key, but there is always an underlying risk that planned funding will not be made available.

Potential contractor arrangements. There are many challenges associated with having separate contractors working on the respective energy and non-energy project scopes. This collaboration could take many forms. In one instance, an ESCO could serve as a subcontractor to a prime contractor delivering non-energy services as part of the SRM project. In this scenario, the agency would not have any privacy with the subcontractor, so they would have to work through the prime contractor. Also, the agency's relationship with the prime contractor would likely be awarded as a construction contract or an operation and maintenance contract, or as a service contract, which could include some construction effort. Those types of contracts would be subject to the FAR, and can generally be in place for only 5 years. This would prevent the agency and the ESCO from benefitting from the partnership of up to a 25-year contract term, which is necessary to deliver substantial energy savings as part of a DER. There are no regulations in place that can bridge the gap of the agency's ability to work with the subcontractor (Fig. 8.3).

There are also challenges if the ESCO is the prime contractor and the agency is trying to incorporate the SRM project or project funding in with the ESCO work. In the combined funding Model #2 illustrated in Fig. 8.4, ESCO is awarded design/build contract for non-energy-related building renovation, and ESPC for energy-related measures. ESCO hires a GC, but provides single point of contact for the government customer.

There has been ongoing discussion to evaluate methods that could be used where an ESCO is in place and has the potential to add value to SRM work. One potential option could be for the ESCO to provide equipment to a prime contractor as

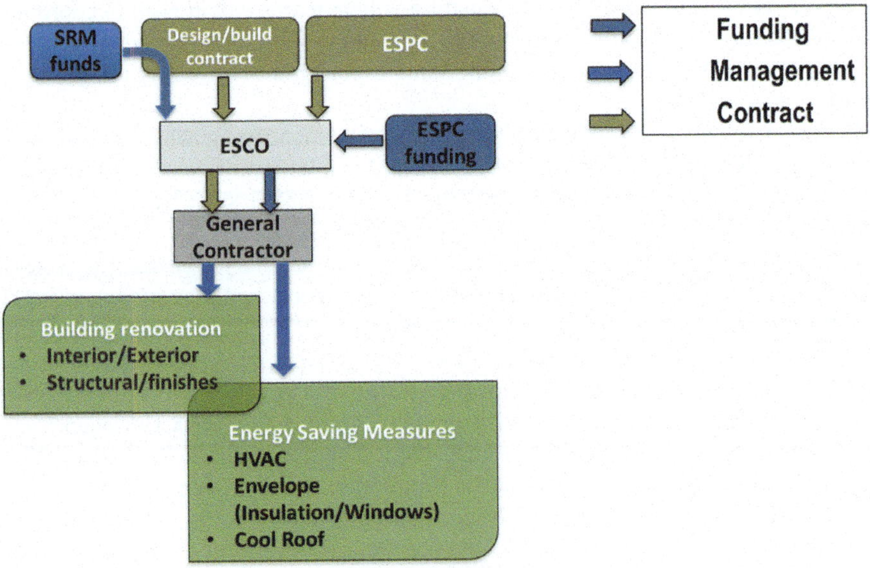

Fig. 8.3 Schematic of the combined funding Model #2

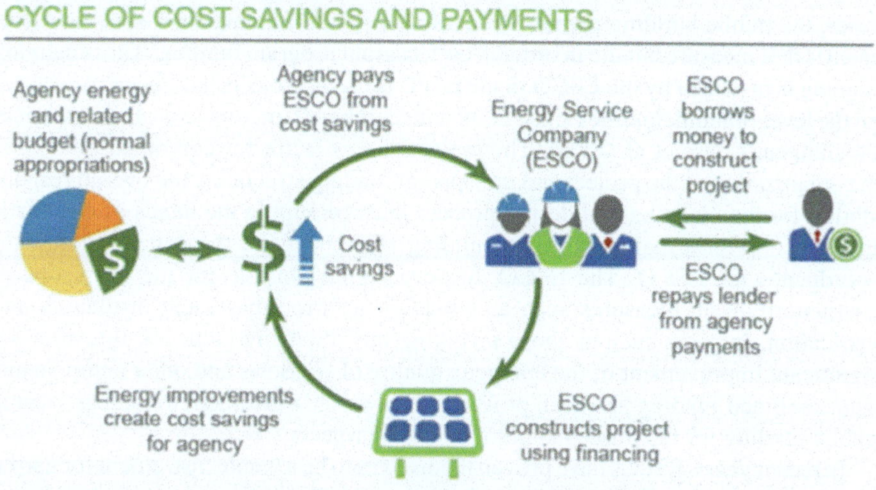

Fig. 8.4 A typical Federal ESPC project structure (DOE 2016c)

government furnished equipment. There are several challenges with how this could transpire, since the SRM contract assumes that the funding covers the entire project (including energy and non-energy scope). The ESCO and an SRM contractor would have to work out the specific arrangements that would allow for this to happen, thereby ensuring that neither contractor performs work outside of the scope of their respective contracts. There could also be challenges during the operation phase of the ESPC if the ESCO alleges that the provided equipment was damaged or not properly installed by the SRM contractor, and that this is the reason that savings are not being realized. So, there are many challenges when separate contractors are hired to perform related energy and non-energy work on an SRM or similar project.

In summary, there are legal issues with how a contract can be structured to comply with 42 USC 8287 and not violate the FAR if appropriated funds are anticipated to be available at the time of contract award. There are privacy of contract issues if the ESCO is a subcontractor to a prime on an SRM project, which would inhibit the agency's ability to accept a comprehensive ESPC project from the prime. There are also issues with an ESCO performing work that is not energy work. Some limited non-energy work could be allowed, but substantial non-energy-related work performed by the ESCO or a subcontractor to the ESCO would not be allowed. So, it is critical that, if there is a potential project that could achieve greater savings using the DER concept, then the team evaluating that project know and understand the procurement rules, and clearly delineate the energy and non-energy scopes to bring the greatest value to the ESPC project.

In some EU countries such as France and Germany, blended funding has mainly been in use for the integration of public grant programs into the funding scheme of ESPC contracts: many national grant programs may be limited to investments provided by public building owners and cannot be accessed by an ESCO. In such cases, the public building owner is providing a partial funding for a specific (dedicated) DER measure bundle in order to collect grant program funding. The remaining funding is provided by the ESCO in the ESPC project. Projects have been carried out on the level of municipalities since 2008 with combined funding (e.g., Projektbericht Plochingen, Lohse et al. 2008). The major success factor for combined funding is the organization of the project and transparent funding structures; the overall responsibility for the planning and design process must be kept in the hands of one party. In projects with a major focus on building refurbishment, the general contractor coordinates the ESCO. The ESCO, however, is responsible for the full scope of energetic relevant measures such as windows, wall/roof/basement insulation, and ventilation. In cases such as those involving the US Army facilities, this allows a significant improvement of the energetic quality of windows and other components. The combined funding has been proved to be a viable instrument to increase scarce public funding for the implementation of DER projects.

In recent years, ESPCs have proven themselves to be a viable instrument for partial refurbishment projects with a major focus on HVAC and energy supply measures. Within the research work and in pilot projects of EBC Annex 61, the scope of ESPCs has increased to include DER. EBC Annex 61 "DER Pilot Projects" reports provide information on holistic refurbishments of the thermal envelope and the NZE projects that have been implemented within advanced DER ESPC business models.

References

Contracting4KMU, R. Lohse, K. Stein, Kompetenzzentrum Contracting, https://www.energiekompetenz-bw.de/contracting/angebote/contracting4kmu/2017

R. Lohse et al., Projektbericht Plochingen (2008). https://www.energiekompetenz-bw.de/contracting/gute-beispiele/

J. Shonder, C. Nasseri, Achieving deeper energy savings in federal energy performance contracts. ASHRAE Trans. **121**, Part 2 (2015)

P. Sweatman, K. Managan, *Financing Energy Efficiency Building Retrofits* (International Policy and Business Model Review, Climate Strategy and Partners and Regulatory Alternatives for Spain, 2010).

A. Zhivov, R. Lohse, J. Shonder, C. Nasseri, H. Staller, O. Moerck, M. Nokkala, Business and technical concepts for deep energy retrofit of public buildings. AT-15–009 . ASHRAE Trans., **121**, Part 2 (2015)

Chapter 9
DER Financing

9.1 Introduction

DER requires cost-effective financing of high investment costs. For public entities, the impact of financing instruments on the balance sheet is relevant for decision-making. In recent years, the bank loans and soft loans have been complemented with a number of attractive financing instruments for the public sector. This chapter describes financing instruments and their usefulness for DER projects from the perspective of public administrators and financiers by the following criteria: flexibility, impact on the public debt-balance, risk mitigation, cost-effectiveness, and options for combining those instruments with other financing tools.

In light of the experience made during the banking crisis of 2008, the EU Capital Requirement Regulation and Directive (CRR/CRD IV) was formulated to apply to credit institutions and investment firms that fall within the scope of the Markets in Financial Instruments Directive (MiFID). Specifically, the risk-weighting under Pillar 1 of the CRD IV requires that the regulatory capital and the liquidity requirements (Liquidity Coverage Requirement Delegated Act) required for any specific asset be in line with the actual risk profile of that asset. The new regulatory capital requirements of Basel III impact EU banks put pressure on the availability of risk capital and on the balance sheets of all financial institutions, and impact energy efficiency investments in all categories. Concern is rising that these new regulations will be blind to environmental targets, and to the long-tail impacts of climate change and the stranded assets that unsustainable and low-resilience investing can create in this context. Obviously, the required capital adequacy ratios may be inappropriate for energy efficiency investments. The accounting regulations for energy efficiency investments neglect to consider the value of inherent multiple benefits, which makes it difficult for financial institutions to allocate investment capital.

A. Zhivov and R. Lohse, *Deep Energy Retrofit—A Guide for Decision Makers*, SpringerBriefs in Applied Sciences and Technology,
https://doi.org/10.1007/978-3-030-66211-0_9

9.1.1 Financial Instruments

Conventional financial instruments that have been used since the oil crises of the 1970s include: grants and subsidies, loans, and tax incentives. Financing is also provided in international funds, either through European institutions such as "European Bank for Reconstruction and Development" and the European Investment Bank, and institutional (mostly environmental or green) funds. The formats used are mostly soft loans and grants distributed by commercial banking institutions. The innovative instruments include EPC (often known as third-party financing) and energy supplier obligations (often known as white certificates). The following definitions of financial instruments illustrated in Fig. 9.1 and their function must be considered:

- Subsidies are handed out to reduce the investment costs of equipment and installations over a certain period of time, i.e., broadening the market approach of a quasi-mature product.

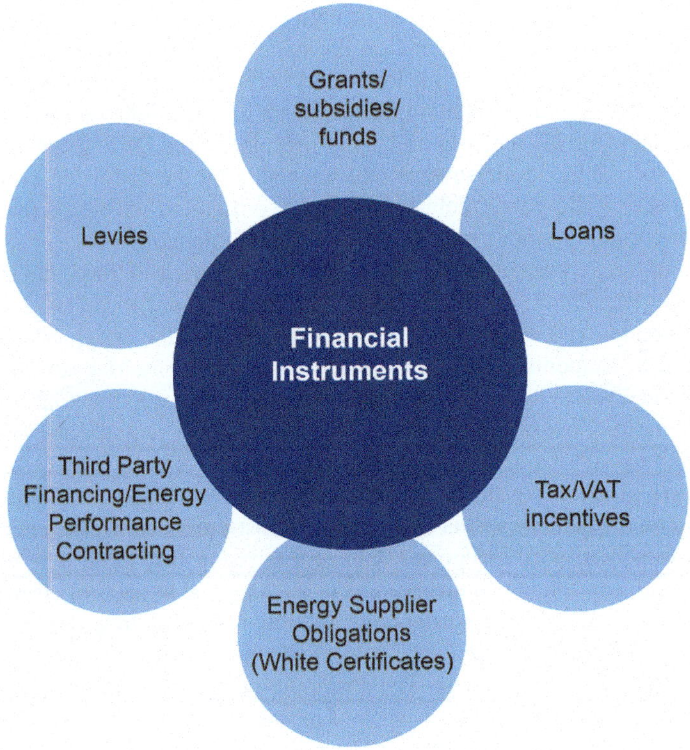

Fig. 9.1 Types of financial instruments that may be available for energy performance improvement of buildings

- Grants are targeted at the end consumers such as households, industrial entities, etc. to pay for a part of the incremental costs of introducing energy-efficient processes in the market—such as enhanced building insulation.
- Grants or subsidies may be financed directly through the state or local authority budget, or through hypothecated taxes (also known as ring-fenced or ear-marked tax).

A handful of criteria are considered when decision is made on financing DER project in the public sector:

- Flexibility of time schedule for the payment of the credit amount to the debtor.
- *Flexibility to increase (or decrease) investment cost totals and expand the borrowed amount of money.*
- *Relevance for the debt-balance of the public building owner.* Countries with a relatively strict austerity policy for the public budget law will have to consider how the financing tool will appear in the public accounting system—as a liability or debt. The assumption is that, in most cases, the credit worthiness of a public entity will, to a certain degree, be related to the same criteria that determine the investment grade of public or in the United States and the UK, public and municipal bonds (Shonder 2014). In other European countries, the re-funding of the public sector is usually provided by special public loan programs such as the "Kommunalkredit" in Germany. The public sector is supervised by public finance and debt control systems, which are located in the ministries of interior (municipalities), the Federal ministry of finance (Federal sector), and which refer to the annual public cash flow budget that each public entity must provide. In this assessment, the creditworthiness is related to the balance of public income (taxes, etc.) and to the liabilities and obligations. Under these premises, loans for public entities are provided as "blank" credits, without securities.
- *Inherited risks and the specific risks of DER projects.* Which are the risks of a financial instrument, where are these risks allocated, and which mechanisms does the instrument provide to reduce the risks? Long-term DER financing agreements bear certain risks such as increasing interest rates, and prepayment penalties that can have a significant impact on the NPV of the project. Also, the performance risk of the DER project must be considered here. From the perspective of the creditor, the normal assessment of the creditworthiness of the obligor cannot be predicted over a term of more than 15 years. The previously mentioned risks of delays and shifting investment cost total are also considered here.
- *Cost-effectiveness of the financial instrument.* The cost-effectiveness of a financial tool is related to the transaction costs necessary to prepare the financing agreement. In many financing instruments, the transaction costs of DER projects are still above average as the number and size of such projects are still comparably small and the due diligence process cannot refer to numerous well-evaluated reference projects.
- *Combination with soft loan and grant programs.* Many EU countries provide a set of subsidiary tools to encourage DERs directly or indirectly by reducing first and annual costs of the DER investment. Well-known programs such as KfW in Germany help to reduce interest rates and the repayment for refurbishment,

which target more ambitious energy efficiency levels than the minimum energy requirements defined in the national standards. Tools that allow the combination with soft loans and grant programs provide a significant benefit for the public building owner.

9.1.2 Performance-Based Financing Instruments

Energy Efficiency Financial Institutions Group (EEFIG) found 16 different financial instruments used for building energy efficiency in the OECD area and has emphasized the importance of the performance-related financing instruments:

Energy performance contracting.
Energy efficiency investment funds.
Public ESCOs for deep renovation in public buildings.
Energy service agreements.

These instruments combine services with a financing instrument, in which remuneration is mainly related to life-cycle cost-related performance indicators. The business model assumes that investments are preparative measures to facilitate the performance, i.e., energy savings in EPC contracts. These instruments are dedicated to public building owners wishing to pay performance-based remuneration instead of an investment sum. The remuneration covers the capital costs of the investment and essential service costs.

9.2 Appropriated Funds

Most of major renovation projects are funded using appropriated funds available to public/government building owners or funded by the commercial building owner. In the United States, appropriated funding comes from the United States Congress, and represents an allocation of Congressional discretionary funding for agencies to realize their approved budgets. Government agencies often have limited appropriated funds to renovate existing buildings, whether to repair aging infrastructure, update building interiors, plan for disaster preparedness and resilience, or perform energy upgrades. Agencies typically have some funding available for building improvements under programs like DoD's SRM program, but it is not often enough funding to retrofit a significant portion of an agency's portfolio on its own.

9.3 Loan Financing, Credit Lines, Revolving Funds, Preferential Loans

A conventional bank loan is the most common form of DER debt in the public sector. After the cost estimation of the architects, the financing plan is set up and the demand for external funding is defined using the construction time schedule. The bank loan is an agreement to lend a principal sum for a fixed period of time, to be paid back within a defined term; the interest rate is calculated as a percentage of the principal sum per year and other transaction costs. Soft loan programs are disbursed by financial intermediaries such as commercial banks. The loan structure depends on the obligor/creditor and on the type of measures to be financed. Loan terms may vary from 5 to 20 years. Typically, the interest rate will be fixed over a certain period of time and will be capped to a maximum throughout this time period. This allows a reduction of risks and opportunities from the interest rate level. The most common method is "annuity repayment," in which the interest plus principal repayment are a constant value over time.

9.4 Soft Loans/Dedicated Credit Lines

Soft loans are subsidized loan programs with no interest or a below-market rate of interest, or loans made by multinational development banks and government agencies to developing countries that would be unable to borrow at the market rate. Soft loans have lenient terms, such as extended grace periods in which only interest or service charges are due, and interest holidays. Soft loans typically offer longer amortization schedules (in some cases up to 50 years) and lower interest rates than conventional bank loans. A dedicated credit line provides low-interest loans to reduce capital costs.

9.5 Non-Recourse and Recourse Finance—Refinancing of ESCO

In comparison to a loan program, the project finance (also: cash flow funding) does take into account the creditworthiness of the obligor and the transactions in which the project is financed based on its own merits. The financed project is often implemented in a project company. In the public sector, project financing is typically used to finance large-scale mission-related projects such as infrastructure measures, social housing, or similar large projects.

This financing format is used to refinance ESCO investments. Due to the lack of experience and performance data, "normal" DER projects that are not based on an EPC business model are not yet considered to be "revenue producing"; this may change over time with the successful dissemination of existing investor confidence programs.

9.6 Private–Public Partnerships

In comparison with normal bank loans or funds procured through combined business and financing models such as public–private partnerships, EPC, or leasing provides certain funded liabilities. The liability of a public building owner in an EPC project is covered by the cost-saving guarantee provided by the ESCO, which has no impact on the debt threshold. However, from the perspective on an ESCO, the creditworthiness of DER projects is currently in doubt. Data are only available for a few projects; the market for DER and the number of DER projects are still low; the overhead costs for risk and creditworthiness assessment are still high; and no well-proven contracts are available for a DER EPC. Also, ESCOs are often at a disadvantage when they attempt to access public grant programs.

9.7 Forfeiting

Financing a forfeit means basically selling future receivables for a discounted lump sum to a bank (forfeiter), normally on the basis of bills of exchange. This financing instrument must be considered in EPC contracts between an ESCO and a public building agency to decrease the financing costs for the ESCO. Without forfeiting, the ESCO must refinance its investment by a project finance contract with or without bank collateral; the ESCO as an industry company will be rated less reliable than a public entity and will receive a higher interest rate (https://www.entranze-scenario.enerdata.eu/site/) than the public entity. With the ESCO forfeiting the future receivables (provided by the public body) to the bank, the bank will receive these receivables. (Fig. 9.2).

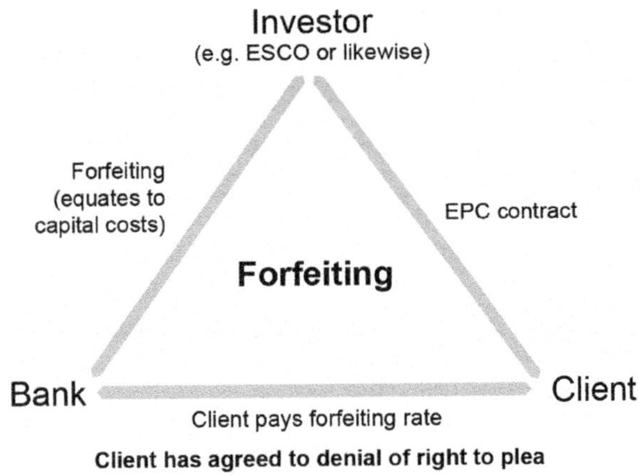

Fig. 9.2 Forfeiting of an EPC project

9.8 Energy Performance Contracting

ESCO financing is a specific format of project financing with a "ring-fenced" project balance sheet that is related to the project costs and incomes. The public building owner appears in the role of contract partner of the ESCO. In most cases, the ESCO provides the pre-finance of the investment. An EPC arrangement is an integrated contract in which a contracting partner (ESCO) designs and implements ECMs with a guaranteed level of energy performance for the duration of the contract. The energy savings are used to repay the upfront investment costs, and, if agreed in EPC contract, the partition of life-cycle costs, after which the contract usually ends.

9.9 Combined Public and Third-Party Financing

Combining a general refurbishment of a building with a DER increases the project's cost-effectiveness and drastically reduces the demand for third-party financing. Since the investment costs for the general refurbishment already have to be provided to start the process, the remaining investment costs are comparably small. Several DER pilot case studies in EBC Annex 61 have demonstrated that the incremental demand for funding to start a DER may be estimated between 10 and 30% of the budget of the general refurbishment project.

Chapter 10
Lessons Learned from Pilot Projects

Some of the concepts developed under the EBC Annex 61 and described in the Technical and Business Guides and their combinations have been tested and further studied as a part of the EBC Annex 61 pilot projects, conducted in Germany, Denmark, the United States, and Estonia. Due to a relatively short duration of the EBC Annex 61 (3 years), these pilot projects had different objectives and resulted in different depth and breadth of information obtained. The objectives varied from testing if DER can be achieved with recommended ECM bundles, cost-effectiveness of DER compared to building a new facility, application of ECMs in combination with RE sources to achieve net or NZE building in a cost-effective way, and demonstrating EPC as a means to finance a DER project. These pilot projects document several "lessons learned" transferrable to future DER projects:

1. All six projects with this objective modeled and adopted a partition of the core bundle ECMs identified in the "DER Technical Guide." Two projects (Darmstadt/DE, Estonia) intended to use Passive House design principles and after evaluation of the first operation period missed that performance goal by 5–10%. Another project (Mannheim/DE), which adopted new building standards that are 40% better than current minimum requirements, is still under construction.
2. Cost-effectiveness compared between pilot projects conducted for the ASHRAE Climate Zone (c.z.) 4a in Europe and the United States shows that lower energy prices in the United States have a negative impact on the cost-effectiveness of a DER refurbishment.
3. Cost-effectiveness is more likely to be achieved in buildings with minimal systems, especially when no or minor air-conditioning and cooling is considered; in the European case studies in c.z. 4a, the buildings had no major air-conditioning and cooling systems in the ex ante status of the project. The DER refurbishment increased the airtightness of the buildings drastically. To maintain required air exchange rates, the European DER case studies implemented ventilation systems including heating and heat recovery units, but only with

A. Zhivov and R. Lohse, *Deep Energy Retrofit—A Guide for Decision Makers*,
SpringerBriefs in Applied Sciences and Technology,
https://doi.org/10.1007/978-3-030-66211-0_10

minor cooling supplies, which added a specific electricity consumption of 5-10 kWh/m^{2}yr compared to the ex ante status of the building. In the United States, the case studies were equipped with ventilation, air-conditioning, and significant cooling supply in both ex post and ex ante of the DER implementation. Thus, the EUIs of comparable European case studies is lower before and after the DER implementation.

4. Cost-effectiveness: In the European case studies, a reduction of the energy baseline by 50% can only be achieved with significant measures on the building envelope and investment costs >200–550 €/m^2. If the value of lower facility energy use in terms of these factors could be monetized and included in the building LCCA, as is suggested in the DER Business Guide, then the additional investments in ECMs could show a positive NPV.
5. DER and NZE: The EEMs (36% energy reduction) for one project (St. Croix) significantly reduced the amount of photovoltaic (PV) needed to make the building net- or near net-zero. NZE, in the definition of a significant reduction of energy consumption from the grid, was not fully achieved in the ex post metered data for technical and organizational reasons. The replicability of the project is related to high electricity prices (0.36 USD/kWh).
6. Demonstrate advanced EPC business models: Energy performance contracting is a business model in which the Energy Service Company (ESCO) invests and implements in energy efficiency and supply measures and is remunerated only with the achieved energy or other life-cycle cost savings. Three projects advanced the scope of the normal EPC business model and applied these successfully to achieve a DER and/or a NZE status: St. Croix, Mannheim, and Silver Spring/New Carrollton. So far, average EPC savings in the United States and Europe have been between 20 and 30%; hence, EPC has not been considered to be an implementing tool for DER or NZE concepts to the building stock. To enable ESCOs to achieve savings >50%, a couple of adjustments had to be carried out.
7. DER EPC specifications necessary: First of all, the building owner has set up a functional specification in which the DER energy-saving target was defined as a major requirement. In the German case, the building owner and the project facilitator carried out a feasibility study and specified cost-effective DER measure bundles.
8. Risk mitigation: Reduction of tendering and bid costs: So far, refurbishment of the thermal envelope and a savings guarantee has not been the technical scope of German ESCOs. To reduce costs and risks for the ESCOs, the tendering specification included some basic architectural requirements that defined the design of some details such as window design, colors, fire protection measures, the minimum air ventilation rate, etc. Also, a modeling tool for the calculation of the energy savings was provided for the ESCOs. This tool has been recalibrated and has been filled with all baseline data, building data, U-values, etc. and allowed the ESCOs to start their calculation on an accurate basis.
9. Risk mitigation—reduction of responsibilities: In a "normal" EPC contract, the ESCO is responsible for the energy savings, the maintenance, and repair over

the contract period. In comparison to the "normal" EPC contract, the implementation of the DER in the German project increased the specific investment costs from 80 to 100 €/m^2 to more than 350 €/m^2. To mitigate the risks for ESCOs, the DER EPC contract allows the ESCO an optimization period of 2 years to finally achieve the savings guarantee. Also, the maintenance of the thermal envelope and the new windows is limited to 5 years (with a total contract period of 16 years). With the adjustments being put in place, ESCOs were able to provide competitive bids. The building owner has been shifting the performance risks for the investment costs, the savings, and the cost-effectiveness of the project to the ESCO.

10. Demonstration of cost-effectiveness of DER when combined with a major renovation: One of the major strategies to improve the cost-effectiveness of DER implementation is to combine DER with major renovations; the assumption that incremental investment costs between energy minimum requirements and DER are cost-effective has been proved so far only on the level of planning and design. However, the first measurement and verification (M&V) period has not been accomplished in most of these case studies. Both projects, which have provided M&V performance data, showed a deviation of 5–10% to the predicted savings and payback periods. Only in a few cases are fully disaggregated investment costs and performance results available; the disaggregation of DER and "normal" refurbishment costs has been done differently in each project and country. The assumption has been positively fulfilled in three European projects: these projects considered all investment costs for the repurposing of the building including the energy-related measures to achieve the minimum energy requirements to be the "normal" refurbishment costs. Here the incremental investment costs to achieve a DER have a payback period of 5–12 years.

Chapter 11
Conclusions

1. Setting up a major renovation in the public building stock involves the allocation of a large amount of scarce public funding. The decision-making process must consider whether it is more cost-effective to refurbish the existing building, or to construct a new building, and if the opportunity should be used to consider EEMs that strive to exceed national minimum requirements.
2. DER can be achieved with a limited core technologies bundle (see Appendix A) readily available on the market. Characteristics of some of these core technology measures described in the DER Technical Guide depend on the technologies available on an individual nation's market, on the minimum requirements of national standards, and on economics (as determined by an LCC analysis). Also, requirements to building envelope-related technologies (e.g., insulation levels, windows, vapor and water barriers, and requirements for building airtightness) depend on specific climate conditions.
3. Characteristics of technologies depend on climate. In hot climate conditions with significant cooling needs, attention shall be paid to reduction of plug loads and advanced HVAC technologies and the use of advanced windows with low solar heat gain coefficients (SHGC). In countries with electricity tariffs beyond 0.15 € or $0.18 (US), the cost-effectiveness can be improved by using energy from renewable energy sources. In heat-dominated climates, the emphasis shall be made on improvement of the building envelope (insulation, airtightness, and remediation of thermal bridges). Eliminate or reduce the need of mechanical cooling when it is not a code requirement and building users can tolerate temporary increases in indoor air temperature (e.g., up to 77 °F [25 °C]).
4. The energy efficiencies and cost-effectiveness of a DER depend on more than the simple characteristics of the core bundle of technologies, how they are implemented, or how they are used. For example, it is important to pay attention to the continuous thermal barrier (no thermal bridges) when building envelope

A. Zhivov and R. Lohse, *Deep Energy Retrofit—A Guide for Decision Makers*, SpringerBriefs in Applied Sciences and Technology,
https://doi.org/10.1007/978-3-030-66211-0_11

insulation is designed and installed, to the continuous air barrier (to achieve required building airtightness), to proper installation of windows in walls, etc.

5. In addition to building energy use reduction, the proper selection, design, and installation of technologies selected for DER result in improvements in indoor air quality, and thermal and visual comfort. A DER usually reduces cold and hot radiation from external walls, prevents drafts created either by air diffusers or by air infiltrating through cracks in the building envelope), improves illumination levels, and eliminates glare through windows.
6. For a DER project to be successful, it is critical to implement a QA process, which, in addition to design, construction, commissioning, and post-occupancy phases, includes formulation of clear and concise documentation of the owner's goals, expectations, and requirements for the renovated building during development of the SOW. Another important component of the QA process is a procurement phase, during which bidders' qualifications, their understanding of the SOW and its requirements, and their previous experience are analyzed.
7. A DER requires a whole-building analysis approach along with an integrative design process. A "whole-building analysis" means that the building is considered as a single, integrated system rather than as a collection of stand-alone systems, such as building envelope, HVAC system, renewable energy system, building operations, etc. The whole-building approach facilitates the identification of synergistic relationships between the component systems.
8. The key to whole-building analysis is the use of an integrated design process. The whole-building analysis differs from a traditional design process in that it brings all relevant disciplines together for an initial charrette-based study of the problem as a whole, based on collaboration and shared information, whereas a more traditional process is based on a linear flow of information passing from one discipline to another.
9. Building systems should be commissioned and adjusted for optimal operation before the project can be handed over to the users/owners and commissioning should be an ongoing activity.
10. The key to making a DER cost-effective is to time the retrofit as part of a major building renovation that already has allocated funds, including those required to meet minimum energy requirements. Since there is an overlap between the funds allocated for the retrofit and those required for the DER, achieving the DER requires only an incremental cost because the DER is evaluated based on a bundle of core technologies, not on individual EEMs. Some "core" technologies (e.g., those related to building envelope insulation, replacement of windows, etc.), which may not be cost-effective when implemented individually, become economically attractive when implemented in a technology bundle. Implementation of these technologies can significantly reduce building heating and cooling loads, and consequently reduce the size and cost of HVAC mechanical equipment, which subsequently results in reduced annual maintenance and insurance costs of these systems.
11. For the DER project LCC analysis, it is important to accurately identify and develop energy and cost model for the pre-renovation baseline, the base case, and

conduct cost–benefit analysis of different DER scenarios. From the cost–benefit analysis provided for each scenario, the decision-maker then selects the one to be further fine-tuned using the LCP approach (see the DER Business Guide). A review of completed refurbishment projects shows that the application of a LCP can improve the cost-effectiveness of DER projects by 5 to 28%.

12. Studies and pilot projects executed around the world by frontrunners indicate that, in addition to the traditional areas of operating cost reduction listed in Sect. 7.6, other bankable cost reduction and income-generating (energy- and non-energy-related) opportunities resulting from DER (listed in Section 7.6) should be considered in an LCCA.
13. When appropriated funds are not available or limited and DER beyond minimum national standards is cost-effective, if possible, use EPC business model as a means to finance the DER project or a combined public and a third-party financing.
14. The behavior of building users has a major influence on the performance of a DER project. Experiences from pilot projects show that misbehavior of building users and facility management staff can lead to a significant underperformance of the DER project, i.e., by inaccurate operation of the building. To ensure optimal performance, users and facility management staff have to be integrated in the preparation and planning phase of the project so they can provide valuable contributions to the design of the concept. Moreover, in such cases, the acceptance of the DER project will be improved. In addition, after the DER has been carried out, the implementation of user training programs is a necessary precondition for a good performance of the DER project.

Appendix A
Technical Aspects of DER

A.1 Building Envelope Technologies

Core technologies and strategies comprising the DER bundle include: insulation of walls, roofs (attic), and slabs; existing windows and doors replacement with high-performance products (including efficient sun shading systems in climates where A/C will be turned on in warm periods); mitigation of thermal bridges; improvement of building airtightness; and vapor control through the building envelope.

A.1.1 Thermal Insulation

Many older buildings undergoing major renovation may have inadequate thermal insulation in the exterior envelope. Existing building insulation types and thermal performance should be evaluated as part of the initial energy audit. The information generated during the energy audit will inform the decisions the design and construction team will need to make in developing the DER strategies and the final DER plan.

The EBC Annex 61 modeling team has optimized building envelope minimum heat transmission (U-values) for wall and roof assemblies and windows thermal requirements through computational modeling performed for representative national climate zones of participating countries. Figure A.1 shows these broad climate zones for the world. More detail for each country and representative city can be found in ASHRAE Standard 169-2013.

Based on results of these studies, the total U-values for wall and roof assemblies required to achieve DER in different climate conditions have been identified and are summarized in Tables A.1 and A.2. These prospective values should be used as starting points for the energy modeling analysis. The final thermal performance

A. Zhivov and R. Lohse, *Deep Energy Retrofit—A Guide for Decision Makers*,
SpringerBriefs in Applied Sciences and Technology,
https://doi.org/10.1007/978-3-030-66211-0

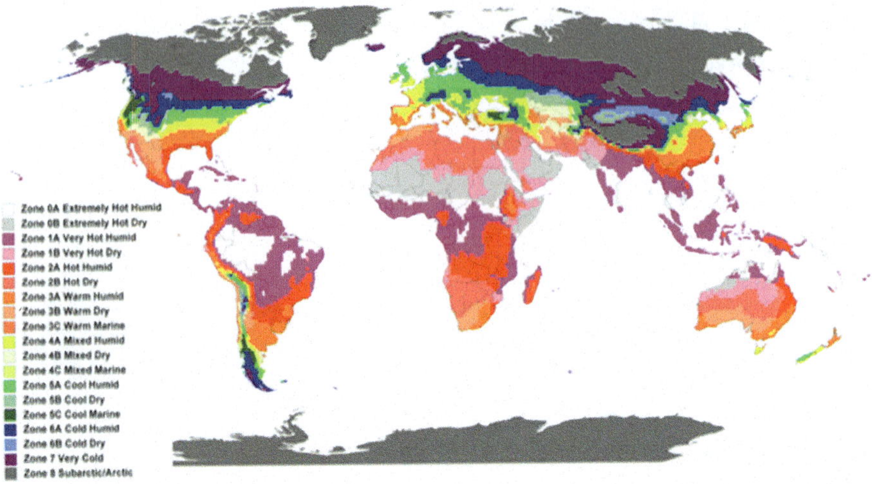

Fig. A.1 Climate zones worldwide (ASHRAE 2013)

Table A.1 Total wall U-value to achieve DER ("c.z." = Climate Zone)

Country	U-value W/(m^2*K) (Btu/[hr*ft^2*°F])	R-value (m^2*K)/W ([hr*ft^2*°F]/Btu)
Austria (c.z. 5A)	0.135 (0.024)	7.4 (42)
c.z.6A	0.24 (0.043)	4.17 (23)
China c.z. 2A	0.96 (0.169)	1.0 (6)
c.z. 3A	0.96 (0.169)	1.0 (6)
c.z. 3C	0.60 (0.106)	1.7 (9)
c.z. 4A	0.48 (0.084)	2.1 (12)
c.z. 7	0.31 (0.054)	3.2 (19)
Denmark (c.z. 5A)	0.15 (0.026)	6.7 (38)
Estonia (c.z. 6A)	0.17 (0.03)	5.9 (33)
Germany (c.z. 5A)	0.17 (0.03)	5.9 (33)
Latvia (c.z. 6A)	0.19 (0.033)	5.3 (30)
UK (c.z. 4A)	0.22 (0.039)	4.5 (26)
5A	0.22 (0.039)	4.5 (26)
USA c.z. 1	0.76 (0.133)	1.3 (8)
c.z. 2	0.38 (0.067)	2.6. (15)
c.z. 3	0.28 (0.050)	3.6 (20)
c.z. 4	0.23 (0.040)	4.3 (25)
c.z. 5	0.19 (0.033)	5.3. (30)
c.z. 6	0.14 (0.025)	7.1. (40)
c.z. 7	0.11 (0.020)	9.1 (50)
c.z. 8	0.11 (0.020)	9.1 (50)

Table A.2 Total roof U-value to achieve DER

Country	Climate zone	U-value W/(m^2*K) (Btu/[hr*ft^2*°F])	R-value (m^2*K)/W ([hr*ft^2*°F]/Btu)
Austria	5A 6A	0.159 (0.028) 0.23 (0.041)	6.3 (36) 4.4 (25)
China	2A 3A 3A 4A 7	0.53 (0.093) 0.53 (0.093) 0.53 (0.093) 0.38 (0.067) 0.30 (0.053)	1.9 (11) 1.9 (11) 1.9 (11) 2.6 (15) 3.3 (19)
Denmark	5A	0.10 (0.018)	1 (57)
Estonia	6A	0.11 (0.02)	9.1 (52)
Germany	5A	0.14 (0.025)	7.1 (40)
Latvia	6A	0.16 (0.029)	6.3 (35)
UK	4A 5A	0.13 (0.023) 0.13 (0.023)	7.7 (44) 7.7 (44)
USA	1 2 3 4 5 6 7 8	0.16 (0.029) 0.14 (0.025) 0.12 (0.022) 0.12 (0.022) 0.11 (0.020) 0.09 (0.0167) 0.09 (0.0154) 0.08 (0.0133)	6.3 (35) 7.1 (40) 8.3 (45) 8.3 (45) 9.1 (50) 11.1 (60) 11.1 (65) 12.5 (75)

values will be determined by results of the detailed modeling performed as a part of the engineering analysis of the DER project.

DER design teams should be aware that nominal wall and roof insulation values such as R-values or U-values will not reflect the actual thermal performance of a finished wall or roof assembly. This is due to the thermal conductivity of framing members, which reduces the overall thermal performance of the finished assemblies. Different national standards provide formulae for de-rating wall and roof insulation depending on the type and spacing of framing members. For example, it is worth nothing that for metal stud walls the overall wall R-value after de-rating is often less than 60% of the nominal cavity insulation R-value. The required U-value of the wall can be achieved by adding:

- Internal insulation.
- Cavity insulation.
- External insulation.
- A combination of insulation strategies listed above.

The major advantage using internal insulation is that it does not change the appearance of the building and that it is relatively cheap to install (compared to external insulation). For historic buildings with protected facades (listed buildings, etc.), internal insulation is the only possibility. It does, however, have some serious drawbacks:

1. It brings a considerable risk of moisture damages, e.g., mold (depending on the climate zone).
2. It leaves a lot of cold bridges that can only be avoided to a certain extent.
3. It will most likely leave some spaces uninsulated.
4. The work causes an inconvenience to occupants.

The primary advantages of external insulation are that it to a large extent eliminates cold bridges and that it is relatively robust with regard to workmanship. The most serious drawbacks are the changed appearance of the building and the price. If a changed appearance can be tolerated—or even is considered a chance to make something new/better—the external insulation is the safest solution with regard to moisture. The price is normally so high that it is hardly ever feasible to make external insulation unless the facade for other reasons needs refurbishment. (A major share of the price goes to scaffolding and surface work.) However, the necessary extent of refurbishment may be reduced as the original facade is being protected by the insulation.

A.1.2 Windows

Fenestration products have a considerable impact on the total amount of building energy usage since windows typically constitute a large percentage of the area of contemporary building facades. This significant amount of fenestration area combined with much higher U-factors than typical opaque wall areas make them disproportionate contributors to building heating loads. Additionally, fenestration decisions made without due consideration for the management of solar gain often result in these same fenestration systems driving the building air-conditioning load. Therefore, any DER project must pay particular attention to window replacement, including area, U-factors, and SHGCs.

Windows also allow daylight into the building and provide occupants visual contact with their surroundings. As such, the optical properties of the selected window replacements can play a key role in defining interior light loads and visual comfort. The local climate of a project and its existing building design constraints (orientation, overhangs, etc.) may further influence proper fenestration decisions. Some climates might imply benefits from transmitted solar energy to offset heating loads, but only if the building design specifically accommodates such beneficial solar gains. However, most existing buildings are not designed to benefit from simple fenestration solar gain, and improperly selected fenestration can easily result in perimeter space overheating, as well as visual and thermal discomfort.

Proper selection of window and fenestration systems is a function of many variables, but all of the decisions start with climate. The climate conditions (combined with building design, type, and occupancy patterns) define peak heating and cooling loads, and also the window energy performance variables to be prioritized.

The window required for subarctic climate (DOE c.z. Zone 8) will certainly be different than that required for very hot and humid climate (DOE c.z. Zone 1)

Table A.3 Minimum window requirements to achieve DER

Country	U-value W/(m^2*K) (Btu/(hr*ft^2*°F))	R-value (m^2*K)/W (hr*ft^2*°F)/Btu	SHGC
Austria			
c.z. 5A	1.09 (0.19)	0.92 (5.3)	0.60
c.z. 6A	1.09 (0.19)	0.92 (5.3)	0.60
China			
c.z. 2A	2.55 (0.45)	0.39 (2.2)	0.48
c.z. 3A	2.55 (0.45)	0.39 (2.2)	0.48
c.z. 3C	2.70 (0.48)	0.37 (2.1)	0.48
c.z. 4A	1.79 (0.32)	0.56 (3.1)	0.68
c.z. 7	1.79 (0.32)	0.56 (3.1)	0.68
Denmark (c.z. 5A)	1.2 (0.21)	0.83 (4.8)	0.63
Estonia (c.z. 6A)	1.1 (0.19)	0.91 (5.3)	0.56
Germany (c.z. 5A)	1.3 (0.23)	0.77 (4.3)	0.55
Latvia (c.z. 6A)	1.2 (0.21)	0.83 (4.8)	0.43
UK			
(c.z. 4A)	1.32 (0.23)	0.76 (4.3)	0.48
c.z. 5A	1.79 (0.32)	0.56 (3.1)	0.68
USA			
c.z. 1	< 2.15 (< 0.38)	> 0.46 (2.6)	< 0.22
c.z. 2	< 1.98 (< 0.35)	> 0.51 (2.9)	< 0.25
c.z. 3	< 1.81 (< 0.32)	> 0.55 (3.1)	< 0.32
c.z. 4	< 1.70 (< 0.30)	> 0.59 (3.3)	< 0.35
c.z. 5	< 1.53 (< 0.27)	> 0.65 (3.7)	< 0.40
c.z. 6	< 1.36 (< 0.24)	> 0.74 (4.2)	< 0.45
c.z. 7	< 1.25 (< 0.22)	> 0.80 (4.5)	0.40– 0.50
c.z. 8	< 1.02 (< 0.18	> 0.98 (5.6)	NR

(Table A.3). For example, in a Climate Zone 8, a window's ability to retain heat inside the building is most important so an extremely low U-factor will likely be prioritized. For Climate Zones 1 or 2, the capacity to block heat gain from the sun will likely be a higher priority.

Table A.3 lists window characteristic determined in modeling studies (Appendix A). These characteristics are minimum requirements that are better than current minimum national standards. They are based on the specific national market conditions and climate-specific considerations, and assume that windows are installed without creating thermal bridges between the frame and the wall. U-values represent average values for the assembly (frame and glazing) and do not reflect variations due to different window sizes and therefore frame to glazing areas ratio. Projects seeking a higher level of energy performance should select windows with lower U-values than those listed in Table A.3.

A.1.3 Historic Windows

In some instances, window replacement may not be feasible or possible. This is the case with windows in historic buildings. The windows are considered an integral part of the historic fabric of the facade and therefore need to remain in place. However, these windows can be renovated to improve energy performance. Weather-stripping can be added or replaced to reduce air infiltration and, depending on the window size and configuration, insulated glazing window units may be installed in place of the existing single-pane glass. In projects where it not possible to perform this type of renovation, either from aesthetic or technical standpoint, an interior storm window may be a viable option. This can have a similar impact on energy performance as a window renovation without the disruption of removing and reinstalling windows. The storm window can include low-E glazing and can also provide enhanced acoustic performance if outside noise reduction is desired.

Note: The energy efficiencies and cost-effectiveness of a DER depend on more than the simple characteristics of the core bundle of technologies, how they are implemented, or how they are used. For example, it is important to pay attention to the continuous thermal barrier (no thermal bridges) when building envelope insulation is designed and installed, to the continuous air barrier (to achieve required building air tightness), to proper installation of windows in walls, etc.

A.1.4 Thermal Bridges

A thermal bridge is an area in an insulated construction that has a significantly poorer degree of insulation than the construction as an average. Thermal bridges are characterized by multi-dimensional heat flows. Figure A.2 shows thermographic images that reveal thermal bridges.

The magnitude of thermal bridges depends on many aspects. Typically, the more insulation a construction has, the greater the relative importance of the thermal bridges. Previous studies show that thermal bridges can increase the total transmission loss from buildings by 14-50%, depending on the extent of the thermal bridges and insulation level of the building envelope in general. This means that thermal bridges are most important to address in new buildings with high insulation levels or existing buildings undergoing DER where insulation levels are increased significantly.

In addition to increasing the transmission of heat loss, thermal bridges also lead to lower internal surface temperatures for constructions, which can lead to

- Poor indoor climate (drafts).
- Contamination of surfaces (dust condensation).
- Moisture damage (mold growth, fungi, etc.).

Thermal bridges typically occur in building joints where different constructions are joined together, i.e., windows or doors in walls. Issues related to thermal bridges,

Fig. A.2 Examples of thermal bridges: parapets, doors, slabs

national requirements to thermal bridges with renovation projects, and the use of infrared (IR) thermography to identify thermal bridging are addressed in (EBC Annex 61 2017).

A.1.5 Improved Building Airtightness

Uncontrolled air transfer through enclosures markedly increases the energy required to heat, cool, control humidity, and regulate indoor climate conditions in buildings. Investigations into building enclosure problems indicate that air leakage is a leading cause of moisture problems (Anis 2001; Zhivov et al. 2014). These include problems of mold, moisture penetration, and durability, especially in intersections between exterior walls, roofs and windows, excessive rain penetration into wall cavities, unstable indoor temperature, and humidity profiles. In colder climates, buildings with insufficient airtightness suffer from moisture-related construction failures, ice damming in roof eaves, icicles on exterior facades, and spalling of masonry. In hot humid climates, infiltrating air in combination with insufficient construction

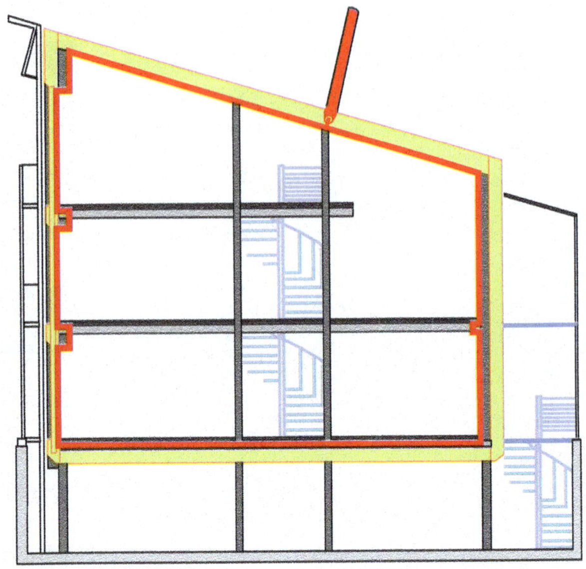

Fig. A.3 Air barrier concept illustrated using the "rule of the drawing pencil" that ensures a continuous air barrier boundary in the contract drawings for a project

thermal bridges causes mold due to condensation on cold air-conditioned surfaces. It is important to clearly identify all air barrier components of each envelope assembly on construction documents and detail the joints, interconnections, and penetrations of the air barrier components (Anis 2001).

The concept of air barrier design that requires one airtight layer all around the whole conditioned (heated, cooled, and humidity controlled) volume may be illustrated using the "rule of drawing pencil" (see Fig. A.3).

In existing buildings, airtightness is often poor due to imperfect joints between the panels, blocks, building components especially around windows, and services penetrating the wall. It is important to address building airtightness with all major renovation projects. If the scope of work of the major renovation projects includes gutting the building, the air barrier concept can be similar to the one used with new construction. In some cases, increasing building airtightness can account for 10 to 40% of the total energy savings.

Table A.4 lists requirements for building airtightness, which differ in different countries (Zhivov et al. 2014) and which are included into the DER energy-saving strategies bundle. Existing buildings undergoing major renovations, especially those located in cold or hot and humid climates, should be sealed to the same standard as new construction if construction details allow for this.

The level of airtightness that can be achieved during the building renovation and approaches to be used depend on the overall scope of work of the project. Buildings undergoing major renovations are typically evacuated and gutted; scaffolding is installed; windows are scheduled for replacement; building envelope insulation is replaced and/or upgraded; and in many cases, roofs are replaced or repaired; most of mechanical, electrical lighting, and energy conversion systems along with connecting ducts, pipes, and wires will be replaced, which requires opening of vertical chases.

Table A.4 Airtightness best practice requirements to achieve DER

Country	Source	Requirement[a]	cfm/ft^2 @ 75Pa*
Estonia	Ordinance No. 58. RT I, 09.06.2015, 21, 2015	≤6 m^3/(h·m^2) @ 50 Pa for renovation ≤3 m^3/(h·m^2) @ 50 Pa for new construction	0.42 cfm/sq ft 0.21 cfm/sq ft
Austria	OIB RL 6, 2011 for buildings with mechanical ventilation	1.5 1/h at 50 Pa	0.28 cfm/sq ft
Denmark	Danish Building Regulations BR10 (2010)	1.5 1/h at 50 Pa	0.28 cfm/sq ft
Germany	DIN 4108-2	1.5 1/h at 50 Pa	0.28 cfm/sq ft
USA	USACE ECB for all buildings (HQUSACE 2010), ASHRAE Standard 189.1-2011, 2013 Supplement, ASHRAE Standard 189.1.–2013 Supplement, ASHRAE Standard 90.1 – 2013		0.25 cfm/sq ft
	USACE HP Buildings and DER proposed requirement		0.15 cfm/sq ft
Latvia	Latvian Construction Standard LBN 002-01 for buildings with mechanical ventilation	2 m^3/(m^{2}h) at 50 Pa	0.14 cfm/sq ft
UK	ATTMA-TSL2	2 m^3/h/m^2 at 50 Pa	0.14 cfm/sq ft
CAN	R-2000	1 sq in. EqLA @10 Pa /100 sq ft	0.13 cfm/sq ft
Germany	Passive House Std	0.6 1/h at 50 Pa	0.11 cfm/sq ft
Sweden	FEBY 12 Std	1.08 m^3/h/m^2 at 50 Pa	0.08 cfm/sq ft

[a]Based on example for four-story building, 120 x 110 ft, n=0.65. Note that values are expressed in the units used in the subject country's national standards

Therefore, with regard to the air barrier, many of major renovation projects can be treated similarly to new construction. In projects with a limited scope of work that does not include gutting of the whole building, there may be only a limited opportunity to improve the building envelope and vertical chases.

A.1.6 Moisture Control

There are flows of heat, air, and water (liquid, capillary, and vapor) through building envelope assemblies. DER Technical Guide (EBC Annex 61 2017) discusses and provides specific recommendations for reduction of heat flow and air flow through assemblies for energy savings. Liquid water flow depends on proper detailing for the exposure. Capillary water flow occurs in all sorptive materials, and its impact is minimal except in materials with ground contact.

Building materials are installed in assemblies for all of the reasons above, as well as for structure and aesthetics. They all have an impact on vapor flow, and that impact depends on their permeance, or their openness to the flow of water vapor under diffusion or the flow driven by vapor pressure differences. Water vapor control consists of ensuring that the permeances of the materials or components in an assembly are such that they do not lead to an unsafe accumulation of water within the building assembly. The requirements for vapor control in above-grade walls and roofs shall comply with ASHRAE Standard 160.

Water management in building envelope assemblies consists of managing liquid water flow, capillary flow, and vapor flow (diffusion). In above-grade assemblies, liquid water flow control is a matter of good design and detailing, for loads that are small and can be easily anticipated. Capillary control is a minor matter where there are no liquid sources.

Water management in foundations consists almost entirely of liquid water management and capillary water management. Vapor management plays a minor role. Liquid water management rarely relies on a single element, but instead includes several elements in series, each of which may be expected to be less than perfect. Methods of liquid water management are discussed in Section 10 of the DER Technical Guide (EBC Annex 61 2017).

A.2 Lighting Systems

Lighting accounts for almost 32% of the energy used in commercial buildings. Related energy codes are becoming more rigorous as the need to reduce energy consumption increases. Since reduction in lighting energy consumption can significantly affect a building's energy performance, lighting is a practical target. Many lighting solutions are simple and easy-to-implement, others more complex; many can yield substantial results. Advanced lighting systems should be considered in all renovation projects of Federal and public facilities.

A number of lighting technologies have been available for decades, but were not often implemented in Federal and public facilities due to either budgetary constraints, lack of guidance, undocumented results, or other application issues. Other technologies in the lighting field are emerging with potential for even greater energy savings if used in the right applications.

When considering energy retrofits, the following basic principles should be considered:

- Provide appropriate illuminance levels without over-lighting.
- Use efficient lamps, ballasts, and luminaires.
- Reduce electric lighting usage with controls.
- Energy-saving lighting design tactics that help create visually comfortable, effective, and efficient lighted environments:
  - Optimize architecture to provide daylight in frequently occupied spaces.

- Apply light-colored (high-reflectance) surface finishes.
- Cluster similar tasks to improve lighting system energy efficiency.
- Locate luminaires close to tasks that require higher illuminance.
- Use LED luminaires predominately.
- Use high-efficiency ballasts with appropriate ballast factors.
- Use high-efficacy versions of lamps.
- Illuminate walls and ceilings to increase perception of brightness.
- Use task lighting in general office areas and locations where additional illumination is needed for detailed tasks.
- Use daylight responsive lighting controls in frequently occupied spaces with daylight access.
- Use occupancy sensors in spaces without daylight access.
- Control lighting with astronomic time clocks for building-wide energy conservation.

There are numerous international guidelines for lighting systems retrofits (e.g., CIBSE 1993, USACE 2013, DS/EN 12464-1:2011, ZVEI 2005). The USACE Lighting Design Guide (USACE 2013) that has been adopted by the EBC Annex 61 provides best practice guidance for lighting strategies in different building types and spaces along with illumination levels and maximum lighting power density (LPD).

A.3 HVAC Equipment and Systems

Mechanical HVAC systems are designed to provide thermal comfort and indoor air quality. The type of system, its efficiency, and its components and control strategies should be selected and designed to minimize energy consumption and to maximize comfort throughout the range of operating conditions.

When building heating, cooling, and electrical loads are significantly reduced, the importance of selecting one type of heating and cooling system over another diminishes.

However, a few aspects must addressed to achieve DER:

- When replacing HVAC systems with new ones, use high-performance motors, fans, furnaces, chillers, boilers, etc. according to the most stringent current national standard and requirements to energy systems.
- Separating systems for ventilation, make-up air, humidity control, and building pressurization from systems providing temperature control can be an effective means of reducing energy use, downsizing mechanical equipment, and improving systems controllability. Use DOASs.
- Use well-sealed and insulated ducts and insulated hot water and chilled water pipes.
- Use energy recovery from exhaust air to preheat and pre-cool outdoor air supplied by the DOAS.

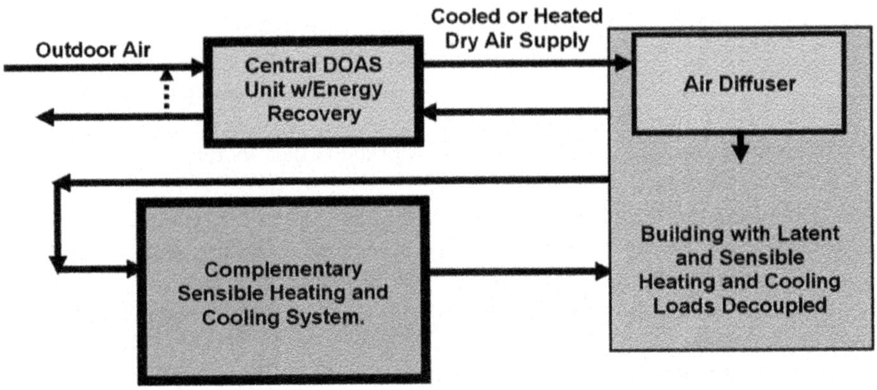

Fig. A.4 DOAS schematic

Requirements for mechanical equipment. Mechanical equipment should be properly sized and meet the efficiency requirements specified in national standards listed in Table A.4 or better. Equipment efficiency should be sized based on optimal performance at part load versus peak its performance. The most important efficiency aspect of HVAC performance is the overall efficiency of the whole system, not just the efficiencies of its components.

One common misconception is that making a building's airtight will result in inferior indoor air quality. Including a DOAS (Figure A.4) into a "core technologies bundle" ensures that building spaces receive sufficient outside air in amounts that are not affected by the building temperature control strategies. A DOAS delivers 100% outside air (OA) to each individual space in the building via its own duct system (PIER 2009) for ventilation, make-up air, humidity control, and building pressurization. Heating and cooling of building spaces are provided by a separately controlled system.

As a general rule, a DOAS operates at constant volume. For most applications, the DOAS cannot meet all of the thermal loads in the space by itself; it requires a parallel system to accommodate any sensible and latent loads the DOAS cannot accommodate.

DOAS airflow rates generally are dictated by

- Indoor air quality needs (based on national standards).
- Latent load (humidity control needs).
- Make-up air for bathroom and kitchen exhausts (when needed).
- Building pressurization to prevent infiltration, which helps to reduce heating/cooling and moisture loads.

Energy recovery. Energy recovery is an important component of the DOAS that recycles energy contained in the air normally exhausted from the building to preheat or pre-cool the incoming ventilation air. In many existing buildings, heating and cooling the ventilation air account for up to 50% of the total HVAC system load.

In such older designs, the ventilation air is treated with conventional components such as an oil or gas burner, hot and chilled water coils, and DX coils. A typical energy recovery process is three to six times more efficient at treating ventilation air compared to a conventional component. To characterize component efficiency, a value called "recovery efficiency ratio" (RER) has been developed, which is the rate of energy recovered relative to the power supplied to perform that work. Similar to "energy efficiency ratio" used in chillers and unitary equipment, RER allows designers to model efficiency gains in whole systems.

The performance of the energy recovery component is typically measured by its effectiveness and pressure drop. Energy transfer effectiveness is defined as the actual energy or moisture recovered to the maximum possible between the exhaust and supply air streams. The energy transfer effectiveness can vary between 15 and 85% with a typical value of 75% for cross-flow energy wheel arrangement (Harvey 2006). It is important to consider the reduced ventilation load and its impact on the primary heating and cooling requirements of the system. Oversizing the system may result in excess energy consumption and may compromise the device's ability to maintain space conditions. Proper system sizing will result in optimized comfort control, energy savings, and minimized capital expense.

Appendix B
Product Delivery Quality Assurance Process

A DER QA process should not be considered a stand-alone process outside of the conventional activities found in all well-managed design and construction projects. Instead, the DER QA process should inform and be incorporated into the relevant phases of project quality control for each project phase or major project activity. There will be overlap between the DER QA process requirements and the typical project QA procedures. To avoid redundancy, duplication of effort, and potential conflicts, a DER QA process should be harmonized and blended into the project QA processes in a seamless fashion.

Examples of conventional project QA activities include SOW and Owner's Project Requirements (OPR) reviews, Charrette Reports, Design Reviews, Contract Document Reviews, Bid Analysis, Commissioning processes, Construction Quality Control, and Post-Occupancy Analysis. DER QA process principles and requirements should be incorporated into each of these processes.

The QA process can be considered as an eight-phase operation, one for each of the basic phases for all construction projects.

The eight phases of construction projects can be viewed as follows:

1. Predesign/Programming.
2. Design Procurement.
3. Design.
4. Construction Procurement.
5. Preconstruction.
6. Construction.
7. Acceptance.
8. Post-occupancy Evaluation.

In addition to these eight phases, there is the commissioning process, which cuts across all of these phases. Not every construction project will be recognized to fall into these specific defined steps, however nearly every project contains these steps in one form or another. The one phase that is least often accomplished formally is the last post-occupancy evaluation.

A. Zhivov and R. Lohse, *Deep Energy Retrofit—A Guide for Decision Makers*,
SpringerBriefs in Applied Sciences and Technology,
https://doi.org/10.1007/978-3-030-66211-0

The following provides a brief summary of each of the eight phases.

1. **Predesign/Programming**—Development of SOW /OPR must provide a clear and concise documentation of the owner's goals, expectations and measurable performance criteria, a cost considerations and benchmarks, and a success criteria to be obtained by the DER for a building renovation. The SOW or OPR shall be used throughout the project delivery to provide an informed baseline for the renovation and focus for design development and for validating building' energy and environmental performance. Based in part on this document, bidders for both design and construction services will be selected on demonstrating their ability to deliver a DER renovation project that will accomplish the goals of the owner as defined in the project's SOW or OPR. Use quality-based sampling for verification of each activity or task to determine how well it meets or relates to the OPR in the Predesign Phase. This includes programming documents, defined scope-of-design services, special reports and workshop outcomes, and other activities in the Predesign Phase. The determination of type procurement process to be used will help frame how the design and construction is selected and the amount of funding required for each phase of the project. In some cases, the owner may elect to obtain assistance in writing the SOW and/or OPR, and in evaluating the proposals before award at future phases of the project. Owners with little experience in building projects, in particular, should consider seeking guidance.
2. **Design Procurement**—The purpose of this phase is to procure the services of a designer who the owner determines will be well suited to provide professional leadership and the design and technical services necessary for the project. The designer must demonstrate a clear understanding of the owner's project requirements, as established by the owners SOW and OPR during the Predesign Phase of the project. In addition, the designer must demonstrate previous design and analytic experience, ability to successfully coordinate different design disciplines, and the ability to deliver the deep energy renovations that meets the SOW/OPR criteria.

   The procurement method for establishing the project contractual basis may be different from project to project. However, all parties should be experienced in the specific project delivery process selected, should demonstrate a clear understanding of the process, and should be comfortable with the approach. Proposals for design services shall be solicited in explicit and clear language. Include a statement regarding design professional commissioning responsibilities and scope in the request for design services. In some cases, the owner may elect to obtain assistance in writing the SOW and/or OPR, and in evaluating the proposals before award. Owners with little experience in building projects, in particular, should consider seeking guidance.
3. **Design**—The design phase starts with concept development by the designer and continues through the completion of documents for bidding or negotiating. During the design phase the following is established: the appearance, configuration, basic system selections, terminations, materials, performance criteria, and interface conditions with other building systems. There is a set of procedures

that the designer must follow at this time to make sure the exterior enclosure systems are appropriately considered, designed, specified, and drawn to attain an enclosure that performs properly, and in compliance with the OPR. Assign operations and maintenance personal to participate in the design phase coordination meetings. The designer must identify how any work excluded to meet budget constraints will affect the OPR and adjust evaluation criteria accordingly.

4. **Construction Procurement**—This should include analysis of construction bidder's qualifications, their understanding of the OPR, previous construction and validation of experience and ability to coordinate different construction trades, performance to meet established schedule and budget and deliver the deep energy renovations in compliance with the SOW/OPR. Bidders should provide evidence of experience installing any type of specialty system that is not commonly used on all types of construction projects.
5. **Preconstruction**—As with the procurement phase, there are provisions the designer can build into the documents, including drawings, technical specifications, and front-end documents for the Preconstruction Phase, the exterior enclosure systems, and other DER provisions.
   The Preconstruction Phase covers the activities between award of the contract and delivery of materials, products, and systems to the building site. This includes final design and engineering, completion of mock-up construction and testing necessary before production, and fabrication and delivery of materials and systems for incorporation into the building.
6. **Construction**—When implementing quality assurance procedures (QAP) during the construction phase, it may be necessary to explain the intent and process of the QAP to all parties. Many site representatives, manufacturers, and tradesmen may not be familiar with the process and may not understand the goals and objectives. However, if you can successfully explain the program and gain their active support and participation, the results can be impressive. Assign operations and maintenance personal to participate in the construction phase coordination meetings.
   During the construction phase of the project, the previous efforts of the owner, designers, consultants, fabricators, material suppliers, and contractors are brought together for the true test. Will it work, will it fit, will it look right, will it perform properly, and can the work be completed on time? If the project team has performed their jobs correctly, and if they have communicated properly, and paid attention to the details, the answer will be yes.
   In-progress testing and inspection of the constructed work, as it occurs, is one of the primary tools of the owner to assure compliance with the project requirements and the SOW/OPR for deep retrofit work.
7. **Acceptance**—Acceptance can be considered an ongoing process applicable to any or all of the phases indicated for a DER, including post-DER occupancy and use of the facility. It can be applied in different forms for different projects.
   Acceptance should include a specific and predetermined approach that is included within the SOW/OPR, and the construction contract documents. Testing and

inspection of the completed work is often the last opportunity to assure compliance with contract documents and performance levels of the various systems before the building is accepted on behalf of the owner.

8. **Post-occupancy Evaluation**—Within the warranty period, key DER project performance should be evaluated to determine whether primary project goals related to energy reduction and system operation are being met. The post-occupancy evaluation should be conducted by members of the original QA process team or a third-party commissioning specialist separately contracted by the building owner.

   Post-occupancy evaluation can be tailored depending on owner budget and project goals. The primary deliverable of the post-occupancy phase is an ongoing commissioning report that documents key building performance metrics including extrapolated annual EUI, zone temperature set point deviation profiles, and HVAC system trend data summaries. Secondary deliverable is an updated Issues and Resolutions Log. Require in the SOW/RFP that the contractor perform corrective action for issues determined to be unrelated to deviations from design or unforeseen conditions within the warranty period. Additional optional deliverables include seasonal endurance testing, occupant comfort surveys, lessons learned workshop, and DER project document updates based on the Ongoing Commissioning (OCx) Report results.

One of the most critical phases of the QA process is development of SOW/OPR. The SOW/OPR are intended to define the scope, goals, and functional performance requirements of the building undergoing DPR renovation. These documents must address both the design and construction phases of the project, and the expectations of the owner relative to the intended use, occupancy, and service life of the DER aspects of the project as well as building users' and building managers' specific needs. These documents must be prepared in sufficient detail to define criteria for both design and construction services, and will provide the basis for verification of understanding and ability to provide services meeting these goals by potential designers and constructors.

The project SOW/OPR should clearly define each goal of the DER project and must be established with realistic and definable goals and criteria. These goals and criteria must be established with the recognition of the limitations and opportunities of the existing building; the goals must be reasonably accomplishable as demonstrated by concept-level engineering analysis, and in the recognition of limitations of contemporary available design and analytical tools, and construction materials, systems, and products. SOW/OPR, which are anticipated to require advanced analytic, design/engineering or materials, products, or systems beyond those readily attainable in contemporary terms, may require a different and specialized approach beyond the scope of this appendix.

It is important that the SOW and bids include specific energy targets, energy security, and system redundancy requirements to be achieved through the DER. Providing the appropriate level of specificity in the SOW/OPR is the first step in assuring that these requirements become contractually binding and will be attained.

There are currently Standards and Guidelines in place through many standards and governmental organizations providing detailed guidance for development of these documents.

Each project and building will be unique and each will require considerable thought in preparing appropriate and attainable SOW/OPR requirements. However, the list below provides a starting point for discussion and perhaps preparing a first draft of an SCW/OPR for many projects. Since each project and building are unique, this list will serve as a guide, but is not exhaustive or fully applicable for all projects.

The SOW/OPR may be prepared by the owner when they have knowledgeable staff skilled in providing such services. Preparation of the SOW/OPR may also be contracted to consultants specializing in the specific areas of DER work being considered. Preparation of the SOW/OPR should consider that:

- In addition to defining the owner's project requirements and criteria through the SOW/OPR process, the methodology by which the goals and expectations have been established must be cross checked before finalization to confirm that they are both reasonable and attainable within the project definition of scope, budget, and schedule.
- SOW/OPR that are not reasonably attainable and do not confirm to other project constraints will not lead to successful projects. (Table B.1)

Table B.1 Selected energy-related parameters and their targets for DER projects to be Included in the SOW/OPR (for more details, see [EBC Annex 61 2017])

Parameter	Sub-parameter	Criteria	Test procedure
Energy Targets	Site Energy, EUI	Minimum of _____KWh/m^2 yr (kBtu/ft^2 yr), >50% reduction compared to pre-renovation benchmark, but better than minimum national/agency requirement	For the bid submission, the bidder shall present a review of the energy requirements for the project to include site and source energy targets; energy calculation and modeling methodologies; and any conflicts or questions from the scope. The bidder shall provide a preliminary results of modeling analysis using simulation program allowing for monthly analysis along with an initial list of all energy parameters for the project to include operational runtimes, load peaks/schedules, equipment efficiencies/set points/sizes, insulation values, etc. Each entry/value shall be identified as being supplied from the project SOW, a specific design guide, a specific energy standard. Simulation results of the renovated building concept at 65% design using computer-based program such as DOE-2, Energy Plus or similar per the ASHRAE Standard 90.1, Section G2.2.1. When economical and feasible compare with the metering data for one year on beginning of the building normal occupancy
	Primary Energy, EUI	better than minimum national/agency requirement	Building modeling at 100% design using local conversion factors
Maximum thermal conductivity of the building envelope	External wall assembly above grade	Table 4 or better (in EBC Annex 61 2017)	Design review and manufacturers' specifications
	Walls below grade	Table A4 (in EBC Annex 61 2017)	Design review and manufacturers' specifications
	Roof assembly, attic floor assembly	Table 5 (in EBC Annex 61 2017) or better	Design review and manufacturers' specifications
	Slabs above unheated basement	Table A4 (in EBC Annex 61 2017)	Design review and manufacturers' specifications
	Slab-on-grade floors	Table A4 (in EBC Annex 61 2017)	Design review and manufacturers' specifications

(continued)

Table B.1 (continued)

Parameter	Sub-parameter	Criteria	Test procedure
	Window assembly	Table 13 (in EBC Annex 61 2017) or better	Design review and manufacturers' specifications
	Opaque doors	Table A4 (in EBC Annex 61 2017)	Design review and manufacturers' specifications
	Skylight	Table A4 (in EBC Annex 61 2017)	Design review and manufacturers' specifications
Thermal bridges		A thermal bridge (or cold bridge) occurs when a thermally conductive material (such as a metal stud, steel frame or concrete beam, slab or column) penetrates or bypasses the exterior insulation system. Design the building envelope to reduce losses from thermal bridging by applying external continuous insulation and aligning all insulating elements, i.e., the continuous wall insulation, insulated glazing, and insulated doors from top of footing to bottom of roof deck. Pay special attention to wall-to-roof, wall-to-wall, and wall-to-window connections. Wrap insulation around roof overhangs. Disconnect window and door sills from interior construction. Use thermally broken window and door frames. Provide details to eliminate thermal bridges particularly at floor slabs, roof/wall intersections, steel lintels and relief angles, metal through-wall flashings and at building corners	Existence of significant thermal bridges shall be verified using thermograph photography on completion of construction before testing air tightness with a pressure difference between inside and outside close to 0 (without building pressurization) and with no direct solar irradiation 3 hrs before and at the time of the test

(continued)

Table B.1 (continued)

Parameter	Sub-parameter	Criteria	Test procedure
Maximum building envelope airtightness		Design and construct the building envelope with a continuous air barrier to control air leakage into, or out of, the conditioned space that shall meet the requirements of Table 20. Design and construct the building envelope with a continuous air barrier to control air leakage in (or out of) the conditioned space. Clearly identify all air barrier components of each envelope assembly on construction documents and detail the joints, interconnections, and penetrations of the air barrier components. Clearly identify the boundary limits of the building air barriers, and of the zone or zones to be tested for building airtightness on the drawings (b) Trace a continuous plane of air tightness throughout the building envelope and make flexible and seal all moving joints. The air barrier material(s) must have an air permeance not to exceed 0.004 CFM/sq ft at 0.3 iwg (0.02 L/s.m^2 @ 75 Pa) when tested in accordance with ASTM E 2178. Join and seal the air barrier material of each assembly in a flexible manner to the air barrier material of adjacent assemblies, allowing for the relative movement of these assemblies and components (c) Support the air barrier so as to withstand the maximum positive and negative air pressure to be placed on the building without displacement, or damage, and transfer the load to the structure. Seal all penetrations of the air barrier. If any unavoidable penetrations of the air barrier by electrical boxes or conduit, plumbing, and other assemblies are not air tight, make them air tight by sealing the assembly and the interface between the assembly and the air barrier or by extending the air barrier over the assembly. The air barrier must be durable to last the anticipated service life of the assembly. Do not install lighting fixtures with ventilation holes through the air barrier. (d) Compartmentalize spaces under negative pressure such as boiler rooms and provide make-up air for combustion e) Provide a mock-up of the exterior wall showing the connections at wall-to-foundation, wall-to-window, wall-to-door, wall-to-roof, wall to intake/exhaust ducts/pipes/conduits/etc. to ensure continuity of the air barrier	Test the completed building and demonstrate that the air leakage rate of the building envelope does not exceed requirements listed in Table 20 in accordance with ASTM E 779 or ASTM E 1827. Conduct the test in accordance with the Engineer and Construction Bulletin "Building Air Tightness and Air Barrier Continuity Requirements" ECB 2012 -16 and the "US Army Corps of Engineers Air Leakage Test Protocol for Building Envelopes, Version 3, February 21, 2012" (http://www.wbdg.org/references/pa_dod_energy.php)

(continued)

Table B.1 (continued)

Parameter	Sub-parameter	Criteria	Test procedure
Mold and mildew prevention		The DoR shall provide details in the design analysis and design showing steps taken to mitigate the potential growth of mold and mildew in the facility. Perform a wall and/or roof construction moisture analysis to verify appropriate thermal insulation and vapor permeability retardant assemblies to prevent condensation with the wall and/or roof under all foreseeable climate conditions. All gypsum board shall achieve a score of 10, the highest level of performance for mold resistance under the ASTM D 3273 test method. All gypsum board shall be transported, handled, stored, and installed in accordance with the GYPSUM ASSOCIATION – *Guidelines for Prevention of Mold Growth on Gypsum Board* (GA-238-03)	Perform a wall and/or roof construction moisture analysis to verify appropriate thermal insulation and vapor permeability retardant assemblies to prevent condensation with the wall and/or roof under all foreseeable climate conditions. All gypsum board shall achieve a score of 10, the highest level of performance for mold resistance under the ASTM D 3273 test method A
Maximum duct and plenum air leakage		During construction, ductwork shall be tested using the pressure test (EN 15727:2010-10, DW143, ANSI/SMACNA 016-2012) to meet Class CL 2.1 (USA) or Class C (EU) ductwork air leakage requirement. The air leakage limit (liters/second per square meter of duct surface area) of the tested duct section, ΔQ shall not exceed: ΔQ_(duct system section) ≤ 0.003 x Δp^0.65 where Δp is the maximum for design operating conditions ductwork pressure difference	ALL supply, return and exhaust air systems shall be tested for leakage during construction to verify good workmanship and the use of low-leakage components as required to achieve the leakage rate specified by owner in the contract. Per ASHRAE (2016) as a minimum 25% of the system (including 100% of the ducts to be enclosed in chases and other concealed space and ducts installed outdoor), based on the duct surface area, should be tested during construction and another 25%, if any of initial sections fail. If any of the second 25% fails, the entire system should be leakage tested. Sections should be selected randomly by the owner's representative. On the whole system completion, it shall be tested at operating conditions using ASHRAE Standard 215 or EN 12599: 2013-01. Leakage tests should be conducted by an independent party responsible to the owner's representative
Minimum duct insulation		Table 12-8 (in EBC Annex 61 2017)	Design review and manufacturers' specifications
Minimum pipe insulation	Hot water pipe	Table 12-9 (in EBC Annex 61 2017)	Design review and manufacturers' specifications
	Cold/chilled water pipe	Table 12-9 (in EBC Annex 61 2017)	

(continued)

Table B.1 (continued)

Parameter	Sub-parameter	Criteria	Test procedure
	Refrigerant pipe	Table 12-9 (in EBC Annex 61 2017)	
Minimum heat recovery equipment efficiency		Table 12-2 (in EBC Annex 61 2017)	Design review and manufacturers' specifications
Minimum HVAC systems equipment efficiency		Table 12-1 (in EBC Annex 61 2017)	Design review and manufacturers' specifications
Lighting systems	Lighting level	Table 22 for (in EBC Annex 61 2017) specific building type and spaces	Design review, Lighting level measurements on construction completion
	LPD	Appendix H (in EBC Annex 61 2017) for specific building type and spaces	
Electric equipment and appliances	Energy efficiency	ENERGYSTAR	Confirm submittal equipment complies
Thermal comfort	Heating season	**ISO Standard 7730, ASHRAE Standard 55 or similar National Thermal Comfort Standard**	
	Cooling season		
Indoor air quality		ASHRAE Standard 62.1 or similar National Ventilation for Indoor Air Quality Standard	

Appendix C
Economics—Strategies to Improve Cost-Effectiveness of DER

The scope of DER project and its attractiveness to investors depend on the project's cost-effectiveness. The standard method to analyze a project's cost-effectiveness is by performing an LCCA, which accounts for present and future costs of the project. Life-cycle costs typically include the following two categories: investment-related costs and operational costs. Investment-related costs include costs related to planning, design, purchase, and construction as well capital replacement costs, which are usually incurred when replacing major systems or components. An LCCA also typically includes energy use cost and the cost of operation and maintenance.

The cost-effectiveness of a DER concept is steered by at least major means that comprise the investment cost optimization, accounting additional non-energy-related life-cycle cost benefits and considering advanced financing tools and business models. In the following, the major approaches that have been considered in EBC Annex 61 are displayed. The scope of Deep Energy Retrofit project and its attractiveness to decision-makers depend on project's cost-effectiveness. The calculation of the cost-effectiveness in the building sector is carried out at the hand of an LCCA. The life-cycle costs are depicted in a cash flow calculation that considers the relevant LCC on the cost (capital costs and operative costs such as maintenance, energy, operation) and on the revenue side (energy cost reductions, increased rental rates, reduced maintenance and insurance costs, etc.). The capital costs refer to the first investment costs that typically include modeling, design, construction, disposal costs; within the life cycle, the reinvestment of worn out first investment equipment is accounted in the secondary investment costs. Capital costs are calculated by an interest rate (internal return rate or loan interest rate) and the calculation period and are displayed as annuities. The key performance indicator is the NPV, which provides the cumulated costs and revenues over a selected time period and a calculated interest rate on today's present value. A positive NPV is cost-effective, a negative NPV means that costs are not covered by revenues over time period selected. Optimization of the cost-effectiveness can be achieved by using LCP calculations.

A. Zhivov and R. Lohse, *Deep Energy Retrofit—A Guide for Decision Makers*, SpringerBriefs in Applied Sciences and Technology,
https://doi.org/10.1007/978-3-030-66211-0

C.1 Value of DER in and Beyond Operational Cost Savings

The life-cycle costs considered in the public building sector are: (1) capital costs that are related to present value of primary and secondary investment costs and (2) operational costs. To improve the cost-effectiveness of DER requires considering the energy cost reductions initiated from energy savings and additional building life-cycle costs. Studies and pilot projects executed around the world by frontrunners indicate that besides energy cost savings additional cost reduction and benefits related to DER have to be considered in the LCCA. One preliminary requirement for any life-cycle cost–benefit is the accountability from the perspective of a funding institute ("bankability"). The bankability means that the funding institute is able to account the life-cycle cost–benefit at least partial as part of the equity; this is the case when the life-cycle cost–benefit can be transparently measured and verified, i.e., at the hand of calibrated meters and one of the involved parties of a DER project takes responsibility or even provides a guarantee that the life-cycle cost–benefit will be achieved. Compared to the base case, DER may result in the following operating cost savings:

1. Energy use and cost reduction due to improved efficiency of the building and its systems resulting from the implementation of energy efficiency measure bundles in a DER project; the value results from the kWh savings and the variable energy price components.
2. Energy cost reduction from fuel switch such as replacement of fossil by alternative biofuels in a new boiler or combined heat and power supply system; the value results from the kWh provided by the alternative fuels and the price reduction per kWh between both fuel sources.
3. Energy cost reduction due to shaving of energy peaks by load management systems, thermal or electric storages the value results from the peak savings in kWh/h and the peak load pricing per hour.
4. Maintenance cost reduction with replacement of aged, worn-out equipment at the end of its life cycle in the context of a DER project; the value results from the annual maintenance cost that have been spent or should have been allocated to the specific component per year and can contribute another 20-30%.
5. Maintenance cost reduction due to downsizing of mechanical systems with reduced heating and cooling loads.
6. Operation cost reduction resulting from the implementation and operation of advanced building automation systems; the value results from the reduced hours per year for the observation and control of major equipment such as boilers, ventilation systems.
7. Capital cost reduction: Grants, rebates, and other financial subsidies for energy-efficient and sustainable design may be reduced by grant-based one-time payments or reduced interest rates of a dedicated loan program; the value is increasing the average energy savings by 30-50% annually.

8. Reduced investment costs by sizing all equipment and the execution of DER project in one phase rather than in several consecutive steps can contribute additional 5-10%.
9. Reduced costs and time of accommodating churn of employees in flexible and sustainable work spaces (single or multiple time cost reduction) that result from a major renovation; usually the reduced costs in comparison to fixed working stations are accounted.
10. Increased usable space due to downsized and consolidated mechanical equipment or improved thermal comfort; the value results from reduced rental rates from the perspective of the tenant or in additional rental rate incomes for the additional available floor space from the perspective of the landlord.
11. Increased usable space due to thermal insulation and ventilation of additionally created attic floor space in a DER; experience from some best practice projects in Norway show that if a roof refurbishment is combined with the creation of additional attic floor space in a DER the rental income can be increased. The value results from the additional floor space, the additional rental rate or sales prices, the increase in usable floor space (close to insulated external walls and advanced windows, and reduced leakage through the building envelope) by ~10%, which produce a value from energy cost savings comparable to an additional 20-50%.
12. Reduced insurance premiums with building components replacement and improved protection against losses; the value is accounted as an annual cost reduction.
13. Staff-cost-related benefits: in comparison to energy costs that usually account to 5-15 €/m^{2}yr the staff cost value per m^2 and per year is 10-30 times higher in normal office spaces. From research in the last 30 years, it is well known that DER has a positive impact on the indoor climate and the productivity. However, only a handful of projects have been generating an evaluation scheme that helps to monetize these benefits. One approach is the reduced short-term absenteeism due to improved indoor air quality and comfort as reduced staff costs. Also, the improved workers' productivity has been considered in LCCA recently. One of the best practice projects, the Comfortmeter project in Belgium, has installed a remuneration system for the facility management company that enables the building owner to account in the best case up to 30 €/m^{2}yr LCC savings.
14. The additional cost benefits from additional rental rates or sales prices have been summarized by the Rocky Mountain Institute (2015) and are shown in Fig. C.1

C.2 Additional DER Cost Benefits

There is increasingly growing evidence from around the world showing non-energy-related economic implications of sustainable renovation of buildings and specifically DER. In these studies, benefits resulting from DER are linked either to different

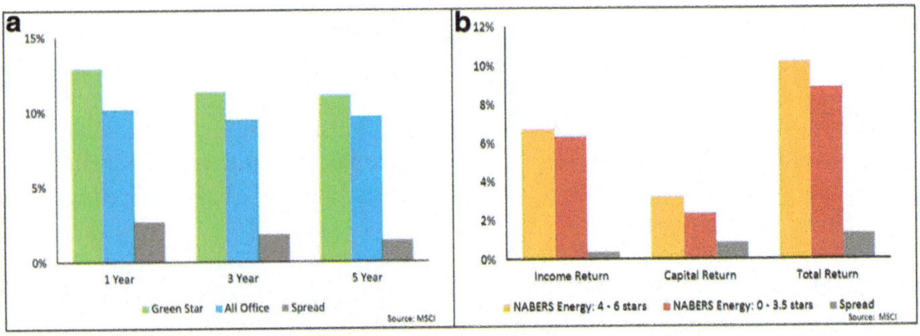

Fig. C.1 Return from Australian CBD office markets to March 2015: a Green Star, b NABERS Energy

green building certification systems, e.g., building certification system LEED® (United States), BREEAM (UK, EU, EFTA member states, EU candidates, as well as the Persian Gulf), Green Star (Australia), CASBEE (Japan), or to only energy focused programs, e.g., ENERGYSTAR or Passive House Standard. While the latter programs' rating is solely focused on energy, different green buildings certification programs in addition to energy- and water-saving attributes include emissions, waste, toxicity, and overall environmental performance criteria designed to reduce the overall impact of the built environment on human health and the natural environment. Many of these sustainable features considered in different certification systems are directly related and required for successful DER project. For example, the LEED certification system has four levels: Certified, Silver, Gold and Platinum, which require 40-49 points for LEED Certified, 50–59 points for LEED Silver, 60-79 points for LEED Gold, and 80 points and above for LEED Platinum. Certain number of points in different categories are necessary just to meet minimum National- or agency-specific sustainability requirement. For example, part experience shows that if a US Army project meets current mandates it can score from 63 to 77 LEED points. In addition to that, DER (> 50% site energy use reduction) allows to score additional points for energy use reduction, which will result in a high overall LEED rating. Therefore, when following original sources of information and refer to different green building certification systems, we understand that buildings having green labels have high energy performance. On the other hand, benefits resulting from technologies implemented as a part of DER go beyond only energy savings and may include elimination of cold/hot radiation from external walls; reduced drafts through cracks in the building envelope; and due to poor performance of the air distribution system, improved ventilation, and therefore better indoor air quality, improved visual comfort, etc., resulting in higher productivity, reduced absenteeism, increased usable building floor area, and other benefits contributing to the building value.

Green building certifications provide a basis for investors to measure and compare properties, a critical foundation for financial analysis (Muldivan 2010). Multiple studies show benefits of DER and different labeling systems certification on their financial performance.

Analysis of 10,000 buildings in the United States containing LEED and/or ENERGYSTAR labeled building and nearby unlabeled buildings reported by (Eichholtz et al. 2010) shows that otherwise identical commercial building with an ENERGYSTAR certification will rent for about 3% more per square foot, and the difference in effective rent is estimated to be about 7%. The increment to the selling price may be as much as 16%. The other paper by Koks, et al. (2012) describes empirical data collected from 374 LEED-certified properties (EBOM) and nearly 600 control properties and shows a 7.1% rental premium for LEED-certified buildings versus non-LEED buildings. When the ENERGYSTAR label is included, they continue to find a significant premium for both ENERGYSTAR and LEED certification.

Evidence provided in Report prepared by Susan Wachter (Undated) shows substantial price and rent premiums that are associated with sustainable buildings in the commercial sector. Studies based on econometrics, combined with current real estate industry data, show that buildings with LEED and Energy Star certification have on average 6% value premium for rents and 15% increase in their sale price.

The Australia Green Property Index Council (MCSI 2015) reports an increase in annualized income return from Green Star-rated offices compared to all reported office buildings 12.9% compared to 6.6%. Offices in high NABERS Energy ratings (4-6 stars) outperform offices with low NABERS Energy ratings (0-3.5 stars), delivering an annualized return of 10.2% (Fig. C.1).

Many sustainable features have multiple impacts on the property value. For example, daylighting can contribute to worker productivity and thereby increase rents. It can also reduce energy costs and thereby reduce operating expenses. Daylighting, if not property implemented, can also result in glare and/or thermal comfort problems (Muldavin 2010). External wall insulation and high-performance windows reduce cold or hot radiation and therefore make people more comfortable at work places located along the perimeter. A DOAS with well-designed air distribution effectively provides sufficient outdoor flow rate, prevents cold drafts, and improves thermal comfort at work stations. The GSA has reported (Muldavin 2010) the following benefits which resulted from best practices of building renovation projects:

- **Reduced absenteeism:** Healthier indoor environments reduce sick building symptoms and absenteeism. A Canadian study revealed that approximately one-third of employees' sick leave can be attributed to symptoms caused by poor indoor air quality. The same study found that communication and social support enabled by open office plans are strong contributors to healthy workplaces and lowered absenteeism. According to a study by Carnegie Mellon University for the DOE, improving indoor air quality and providing natural light reduce illness and stress. The CMU study (Loftness et al. 2003) showed that occupants closer to windows reported fewer health problems. In addition, a survey of three case studies by the Rocky Mountain Institute proved that better lighting and HVAC systems could reduce absenteeism from 15 to 25%.

- **Improved recruitment and retention:** The workplace is a proven factor in hiring and keeping a world-class workforce, resulting in improved recruitment and retention rates and decreasing expenses to replace staff. Knoll reports that a Hay Group study found that half the people planning to leave their current employer were dissatisfied with their workplace, while only one-quarter of those staying were dissatisfied. A study commissioned by the American Society of Interior Designers also found that 51% of employees surveyed said the physical workplace would impact their decision to leave their job.
- **Increased productivity and performance:** Flexible, adaptable work settings allow people to customize their workspace to suit their individual needs, providing improved comfort. When given control over their environment, workers are less distracted and more productive and satisfied with their jobs. They also report fewer complaints to building management. For example, Public Works and Government Services Canada found that when people were given individual ventilation control, the number of trouble calls decreased significantly.
  Healthier, more ergonomic workplaces can also improve performance and reduce expenses. The Occupational Safety and Health Administration (OSHA) reports that repetitive strain injuries caused by poor ergonomic design, including computer use, cost business, and industry as much as $54 billion annually in workers compensation and other costs.
- **Greater flexibility of building services:** Improved flexibility in workplace design reduces the time and expense required for reconfigurations and daily operations and maintenance. The GSA Adaptable Workplace Lab showed that using easily reconfigured furniture can save 90% of reconfiguration costs, and reduce reconfiguration time from days to hours. In another example, the Pennsylvania Department of Environmental Protection reduced average churn costs from $2,500 to $250 per workstation by using more flexible building and furniture systems in their high-performance green buildings.
- **Efficient operations and maintenance.** Innovative workplaces help decrease facility management, operating, and technology expenses. Vivian Loftness et al. at Carnegie Mellon have compiled case studies that show that improved lighting efficiency and control can save up to 40% in total building energy costs.

Deep retrofits can result in measurable, but indirect sources of new operating income. DERs often result in reduced vacancy rates due to greater occupant satisfaction, which the GSA has reported as 27% higher than national averages in post-occupancy surveys for its "green" buildings (Fowler 2008; GSA 2011).

Additionally, insurance companies have recently begun rewarding green buildings with reduced premiums, citing the contention that commissioning and sustainable design reduce "sick building syndrome" claims, and may also reduce damage claims from both human and natural hazards (Nalewaik et al. 2010).

Deep energy retrofits also bring a host of secondary benefits that are more difficult to quantify, but can result in indirect cost savings. These include productivity increases, increases to health and well-being, recruiting advantage and greater employee retention, and property value increases, among others (Bendewald 2015).

While these benefits are challenging to quantify and cannot be counted as finance-able cost savings, they should still be promoted and considered as added benefits in a DER project.

The survey data presented in (Leonardo Academy 2008) collected during 2006-2007 from owners or managers of 23 LEED-EB-certified building allowed to make following key conclusions:

- The costs for LEED-EB implementation and certification varied significantly from building to building. The total costs were a mean of $2.71 per square foot, with a median of $2.31 per square foot. The results did not follow expectations of higher costs for higher certification levels, but this may be due to the very small sample size available.
- In all the categories of operating costs, more than 50% of the LEED-EB buildings had expenses less than the BOMA average for the region. Total expenses per square foot of the LEED-EB buildings were less than the BOMA average for seven of the 11 buildings (64%).
- Total operating expenses in LEED-EB-certified buildings had a median of $6.07 per square foot, 13% less than the $6.97 average for BOMA buildings.

C.3 Value of DER in Public Buildings

The above discussion shows that significant progress has been made in the real estate industry in quantifying and articulating the value of sustainable property investment. Most investors, and many tenants, today understand that sustainable properties can generate health and productivity benefits, recruiting and retention advantages, and reduce risks, but struggle to integrate benefits beyond cost savings into their valuations and underwriting (Value Beyond Cost Savings: How to Underwrite Sustainable Properties).

However, this concept has not yet found a traction in the Public/Government sector. As the result, most of decisions related to the scope of energy and sustainability work under building major renovation project are based on cost-effectiveness resulting from operating cost reduction.

To promote the concept of DER value to the public sector, Fig. C.2 shows a trend of building value change throughout its useful life. Assume that the building has some value **V1** "on books" on construction completion and beginning of its operation. Throughout its operation the building value depreciates and has some residual value **V2** at the end of its useful life, when major renovation is warranted. Throughout the process the value can slightly increase with some minor renovations or drop due to man-made or natural damage to the building, which may require that major renovation is required before the planned end of its useful life. With the major renovation, the value of the building will increase and reaches the **V3,** which probably will be higher than **V1** due to more stringent standard requirements to the building (seismic, thermal comfort, indoor air quality, etc.) compared to those used during its design and construction. If the major renovation is combined with DER and follows

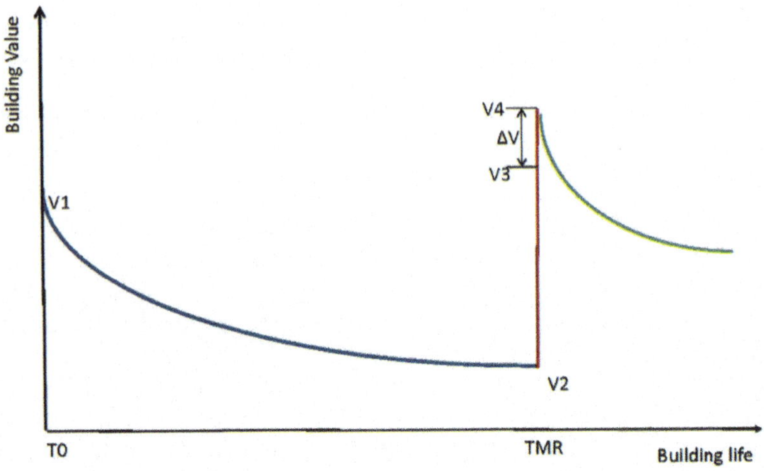

Fig. C.2 Building value over its life

energy and sustainability requirements beyond minimum standard requirements at the time of its renovation, the value of the building will increase to **V4** by ΔV = **V4-V3**. Based on the data available from the private sector studies described above, it is safe to suggest that ΔV may be at least 5%, which when annualized can be used in LCCA in addition to operating cost savings for justification of cost-effectiveness of DER.

C.4 Strategies for Investment Cost Reduction

Selection and optimization of the DER scenario. Modeling of energy scenarios is the first step to optimize the cost–benefit ratio of a DER project: here, different energy scenarios are modeled, resulting in U-values of building envelope, airtightness levels, ventilation, and heat recovery rates. In the next step, investment costs and energy cost savings are estimated and then compared to the base case scenario, which describes the national minimum requirements for a building refurbishment. In addition to that, at least three scenarios should be compared to the base case scenario; these could be new building standard target values adapted to the refurbishment case, a DER scenario of -50% compared to the baseline and a "dream scenario," which could be Passive House, NZE. In this context, the baseline reflects the usage and climate-adjusted average utility data of the recent 2-4 years; the baseline is used for benchmarking of energy use in the building prior to renovation and for the recalibration of the modeling calculations. The result of the modeling will display the most cost-effective scenario. In a next step, this scenario will be fine-tuned and recalculated at the hand of a LCP approach. The LCP is carried out in iterative calculations either

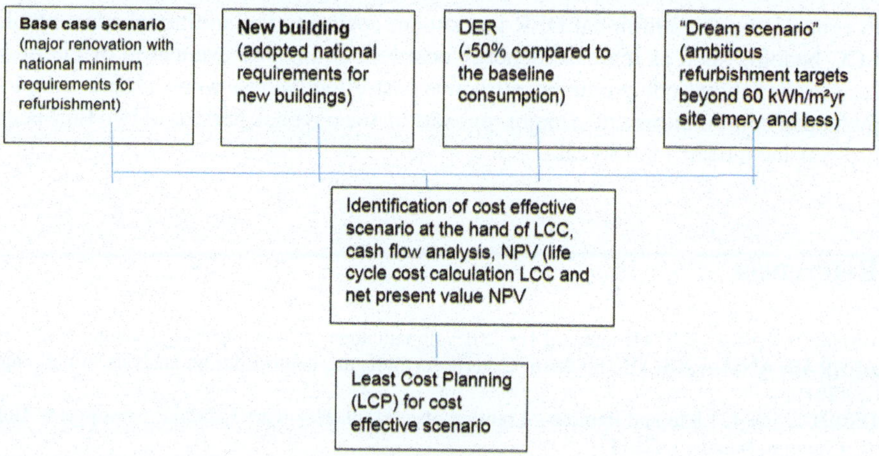

Fig. C.3 Improving cost-effectiveness: Modeling and LCP

with the modeling software or by using specific LCP tools (Jank et al. 2017). It will compare the incremental cost of 1cm of additional wall, roof, and basement insulation with double- /triple-pane windows, ventilation system costs with different heat recovery factors, etc.; the result will be a streamlined cost–benefit approach for the selected refurbishment scenario. A review of accomplished refurbishment project shows that the application of a LCP is not yet "business as usual" and may improve the cost-effectiveness of DER projects by at least 5% and up to 28% (Fig. C.3).

Experience from the pilot case studies shows that the *combination of a general refurbishment and a DER* is one of the most viable ways to improve cost-effectiveness. Major renovation projects include a scope of work with non-energy-related and energy-related investment costs. In a major renovation, the non-energy-related scope of work may include different construction jobs related to changing floor layouts, removing harmful materials, adding fire protection measures, facade concrete renovation, roof sealing (also scaffolds), and construction plant equipment. Major renovation also includes measures that have an impact on the energy balance, e.g., replacement of existing mechanical, lighting, and electrical systems and replacement of windows, ductwork, and plumbing systems. Major renovation with the energy-related scope of work that will meet current minimum standard requirements will be considered to be a base case. While non-energy-related scope of work will remain the same in both the base case and DER scenarios, the energy-related scope of work with DER can include the same items, but using higher efficiency equipment and systems, as well as additional items, i.e., building insulation, improvement of building air tightness, heat recovery, etc. In comparison to the base case, any other more ambitious refurbishment scenario will increase the energy-related investment cost. However, the implementation of DER also creates synergies that allow to downsize heating and cooling systems, boilers and chillers, which will reduce energy-related investment costs. The cost-effectiveness of the DER will have only to consider by the additional DER investment costs and the additional accountable LCC benefits such as increased energy cost reduction. The evaluation of the pilot case studies shows comparatively attractive cost-effectiveness when a DER is implemented in combination with a major renovation: the payback periods of two European projects have been 11–18 years.

References

ASHRAE, ANSI/ASHRAE standard 169-2013 climatic data for building design standards (2013)

ASHRAE, ANSI/ASHRAE 160-2016. Criteria for moisture-control design analysis in buildings (2016)

DS/EN 12464-1, Light and lighting—Lighting of work places—Part 1: Indoor work places. Eur. Committ. Standard. (2011)

EBC Annex 61, Deep energy retrofit—a guide to achieving significant energy use reduction with major renovation projects. Subtask a report (2017). www.iea-annex61.org

P. Eichholtz, K. Nils, J. Quigley, Doing well by doing good? Green office buildings. Am. Econ. Rev. **100**, 2492–2509 (2010, December). http://www.aeaweb.org/articles.php?doi=10.1257/aer.100.5.2492

K.M. Fowler, E.M. Rauch, Assessing green building performance, a post-occupancy evaluation of Twelve GSA buildings, pacific northwest national laboratory (2008, July)

GSA, Green building performance. A post occupancy evaluation of 22 GSA Buildings. White paper (2011, August)

D. Harvey, A handbook on low energy buildings and district energy systems. Earthscan. Sterling, VA (2006)

R. Jank, M. Korb, R. Lohse, Deep energy retrofit optimization model (DEROM) handbook (in German), KEA, Karlsruhe, Germany (2017)

Leonardo Academy, The economics of LEED for existing buildings. Leonardo Academy Inc.(2008, April 21)

MCSI, The property council/IPD. Australia Green Property Index. MSCI.com. (2015)

S. Muldavin, Value beyond cost savings: how to underwrite sustainable properties. The Muldavin Company, Inc. (2010)

USACE, Lighting design guide for low energy buildings—new and retrofits. ERDC (2013)

A. Zhivov, D. Herron, L. Durston, M. Heron, G. Lea, Airtightness in new and retrofitted US army buildings. Int. J. Ventil. ISSN 1473-3315 **12** (4) (2014, March)

ZVEI Manual to DIN EN 12464-1. Lighting of work places—Part 1: Indoor work places, ZVEI (Zentralverband Elektrotechnik und Elektronikindus [German Central Association of Electrical and Electronic Industry]) (2005).

Bibliography

M. Bendewald, D. Miller, S. Muldavin, *How to Calculate and Present Deep Retrofit Value* (RMI.org, A Guide for Investors, 2015)

Bartoldi et al., Energy Service Companies Market in Europe—Status Report 2013, Joint Research Centre, ISPRA, Italy (2013)

CIBSE, Lighting Guide 7: Lighting for Offices. The Chartered Institution of Building Services Engineers (CIBSE) (1998). ISBN: 0900953632

Dobbs Richard, Jeremy Oppenheim, Fraser Thompson, Marcel Brinkman, Marc Zornes, Resource Revolution, McKinsey Global Institute Report (2011). http://www.mckinsey.com/business-functions/sustainability-and-resource-productivity/our-insights/resource-revolution

Entranze H2020 project, (2015) http://www.entranze-scenario.enerdata.eu/site/

Entranze scenario results. http://www.entranze-scenario.enerdata.eu/site/

IEA/OECD, Capturing the Multiple Benefits of Energy Efficiency. IEA/OECD, Paris, France (2014). https://www.iea.org/publications/freepublications/publication/ Capturing the Multiple Benefits of Energy Efficiency .pdf

IEA/OECD, Technology Roadmap Energy Efficient Building Envelopes. IEA OECD, Paris France (2014), https://www.iea.org/publications/freepublications/publication/TechnologyRoadmapEnergyEfficientBuildingEnvelopes.pdf

Kock Nils, Norman G. Miller, Peter Morris, The economics of green retrofits. J. Sustain. Real Estate (JOSRE) **14**(1) (2012)

R. Lohse, Scaling up lighting system refurbishments in public buildings. Report (German) KEA, Karlsruhe, Germany (1998)

B. Metz, O.R. Davidson, P.R. Bosch, R. Dave, L.A. Meyer, *Climate Change 2007* (Cambridge University Press, Cambridge, UK, Mitigation of climate change, 2007)

A. Nalewaik, V. Venters, Costs and benefits of building green. IEEE Eng. Manag. Rev. **38**(2), 77–87 (2010, June)

NEA, Guaranteed Energy Savings Performance (GESP) Contracting Model. http://www.e2singapore.gov.sg/DATA/0/docs/GESP%20Contract/Overview%20of%20the%20GESP%20model.pdf. National Environment Agency, Singapore

PACE, Property Assessed Clean Energy. www.pacenation.us/commercial-pace

A. Zhivov and R. Lohse, *Deep Energy Retrofit—A Guide for Decision Makers*, SpringerBriefs in Applied Sciences and Technology,
https://doi.org/10.1007/978-3-030-66211-0

J. Shonder, R. Lohse et al., *Investors Day 2014 Summary* (KEA, Karlsruhe, Germany, 2015)

M. Ungclaus, C. Carmichael, C. McClurg, M. Simmons, R. Smidt, K. Porst Hydras, S. Conger, A. Zhivov, F. Winter J. Shonder C. Nasseri, Deep Energy Retrofits in Federal Buildings: The Value, Funding Models, and Best Practices (2017). ASHRAE Transactions, Volume 123, Part 1

Volkswohnung, Refurbishment cost data evaluation for triple pane windows. Volkswohnung, Karlsruhe, Germany (2017)

Wachter. Susan, Valuing Energy Efficient Buildings. Report prepared for Energy Efficient Building Hub. http://gislab.wharton.upenn.edu/Papers/Valuing%20Energy%20Efficient%20Buildings.pdf

The manufacturer's authorised representative in the EU is Springer Nature Customer Service Centre GmbH, Europaplatz 3, 69115 Heidelberg, Germany. If you have any concerns regarding our products, please contact ProductSafety@springernature.com

Printed and bound by CPI Group (UK) Ltd, Croydon, CR0 4YY
22/09/2026
02220957-0008